DIGITAL JOURNALISM

DIGITAL JOURNALISM
Emerging Media and the Changing Horizons of Journalism

Edited by
Kevin Kawamoto

ROWMAN & LITTLEFIELD PUBLISHERS, INC.
Lanham • Boulder • New York • Toronto • Oxford

ROWMAN & LITTLEFIELD PUBLISHERS, INC.

Published in the United States of America
by Rowman & Littlefield Publishers, Inc.
A wholly owned subsidiary of The Rowman & Littlefield Publishing Group, Inc.
4501 Forbes Boulevard, Suite 200, Lanham, MD 20706
www.rowmanlittlefield.com

P.O. Box 317, Oxford OX2 9RU, UK

British Library Cataloguing in Publication Information Available

Library of Congress Cataloging-in-Publication Data

Digital journalism : emerging media and the changing horizons of journalism / edited by Kevin Kawamoto.
    p. cm.
Includes bibliographical references and index.
    ISBN 0-7425-2680-1 (cloth : alk. paper) — ISBN 0-7425-2681-X (pbk. : alk. paper)
  1. Electronic journals. I. Kawamoto, Kevin.
    PN4833.D55 2003
    070.4—dc21

                                    2003008617

Printed in the United States of America

♾™ The paper used in this publication meets the minimum requirements of American National Standard for Information Sciences—Permanence of Paper for Printed Library Materials, ANSI/NISO Z39.48-1992.

Contents

Acknowledgments   vii

Preface   ix

1   Digital Journalism: Emerging Media and the Changing
Horizons of Journalism
Kevin Kawamoto   1

2   The History of Online Journalism
David Carlson   31

3   The Meanings and Implications of Convergence
Rich Gordon   57

4   New Technology and News Flows: Journalism and
Crisis Coverage
John V. Pavlik   75

5   Digital Photojournalism
Cheryl Diaz Meyer   91

6   Satellites, the Internet, and Journalism
Adam Clayton Powell III   103

7   Social Movements and the Net: Activist Journalism
Goes Digital
Melissa A. Wall   113

8   Digital Government and an Informatics of Governing:
    Remediating the Relationship between Citizens and
    their Government
    Paul W. Taylor   123

9   Online Medical Communication among Peers:
    The Net and Alternatives to Traditional Journalism
    Patricia Radin   145

Conclusion
    Kevin Kawamoto   167

Multimedia Coverage: A Case Study and Exercise   177

Index   183

About the Contributors   187

Acknowledgments

BRENDA HADENFELDT, acquisitions editor at Rowman & Littlefield, planted the seed for this book and nurtured it through to its fruition. She had the vision to realize that the field of journalism and mass communications would benefit from a book in this format, dealing with a subject that has implications for journalism education and the news industry today—and will probably continue to do so for years to come. It was a great privilege to work with her and the talented authors who graciously agreed to contribute their time, energy, and knowledge to this project. Good writers benefit from working with good editors. Alden Perkins, an associate editor at Rowman & Littlefield and the production editor for this book, and Jen Kelland, the copyeditor, improved this book considerably with their careful and thorough editing. The authors also wish to thank the reviewers who provided incisive, constructive, and informed criticism of the manuscript. We were fortunate to have reviewers with the interests of students and educators at heart. John Klockner, who oversees technology at the University of Washington Department of Communication, was always generous with his time, advice, and support. Thanks, too, to award-winning photographer Cheryl Diaz Meyer, who allowed her photographs to be published in this book, and to Mark Brender, executive director of government affairs and corporate communications, Space Imaging, for granting permission to use two satellite images that appear in chapter 4. With nine authors contributing to this book, it would be impossible to thank everyone who helped us in personal and professional ways during the writing and production process. We must be content to extend a blanket, but heartfelt, thank you. This book is dedicated to the late Dr. Patricia Radin, a contributor to the book, who died unexpectedly while it was in production. She was an educator, an award-winning journalist, and a friend to many people and causes. Her work in health communication will live on through those of us who were inspired by her passion, insight, and intelligence.

Preface

THIS BOOK IS A COMPILATION OF ESSAYS about digital journalism—what it is, where it is going, and why it is important—by a group of educators, journalists, and other media professionals who have helped nurture this growing field for many years. The study of digital journalism (also known by monikers such as online journalism, cyberjournalism, new media journalism, interactive journalism, and multimedia journalism) is rapidly taking root in journalism and communication schools in the United States and abroad.

Digital journalism was not always a welcome addition to the academic curriculum or the news industry. In the early 1990s, some faculty members and industry professionals believed that the Internet and World Wide Web were nothing more than passing fads and therefore not worthy of close attention by serious journalists and journalism educators. Today few would disagree that digital media technologies are an important and enduring feature of the global communications landscape and that they will have as significant an impact (if not more of one) on society as books, newspapers, magazines, radio, and television.

Research from various sources has shown for several years now that more than half of U.S. households have a personal computer. A 2002 survey by the Pew Research Center revealed that nearly two-thirds of Americans report having a home computer and 71 percent use a computer from somewhere at least on an occasional basis.

Thirty-five percent of Americans, according to this study, go online for news at least once a week. This is a dramatic increase from 1995, when only 5 percent went online for news at least once a week (although the growth has slowed in recent years).[1]

Clearly online news is becoming a growing part of Americans' news diet. Research suggests that at this stage the role that online news plays is supplemental, not substitutive. Even those who are heavy users of the Internet for news still look at other media as sources for news.[2] But online news is on that "menu" of sources, and most news organizations have been allocating resources to make content available to their audiences in digital form since the mid-1990s (and some even earlier).

This transformation has led to the concept of digital journalism, which is a difficult concept to define (see chapter 1). It is a new enough concept that there is no absolute agreement about what it is. Some might consider the mere repurposing of content from one traditional medium (e.g., newspapers or television) to a digital medium (the Web) as having sufficiently met the criteria for digital journalism. Others might require that content be specifically designed for a digital medium—at least making use of hypertextuality and interactivity—by journalists trained in digital media "theory" before qualifying as digital journalism. The definition is still evolving, as is the field as an area of study.

CNN Interactive hosts a Web page that explains to its readers how stories end up on the Web site, a process that involves the interaction between news professionals from the cable television side of the operation and other specialists from the online division.[3] It involves a team effort among various producers, editors, and Web specialists. This is typical of many news organizations, whether print or broadcast, in that a symbiotic relationship exists between the traditional journalist who creates content (e.g., news stories) for a traditional medium and the new media division of the organization, which is responsible for making that content accessible to an online audience. But this is only one aspect of digital journalism. There are other news organizations that do not have a traditional media counterpart, but only exist as digital entities. Other news organizations are requiring their journalists to be multimedia competent, so that they no longer pigeonhole themselves as a specific kind of journalist, but rather see themselves as having multiple competencies. These are also aspects of digital journalism.

Is it good to have journalists who are multimedia competent? The jury is still out on that question. One could argue several ways. This book often makes the case that multimedia competency is not only good, but also an inevitable job requirement for journalists. But one could also argue that journalists should not be encumbered by the annoyances of new technologies. It distracts them from their primary responsibility, which is writing and reporting the news. This is one of many questions that should be left for discussion in classrooms and newsrooms. The chapters in this book will help to generate others.

Changing Horizons of Journalism

This book contains ten chapters written by nine experts in their respective subfields of digital journalism. Journalists entering the news industry in the twenty-first century will be exposed to a wide range of knowledge and perspectives from people who have the benefit of a special vantage point because of their experience in the field. The authors of this book represent a diversity of backgrounds: journalism education, print and broadcast media, digital media, computer-assisted research, government information services, media management, photojournalism, media and social movements, health communication, and other specialty areas. As a group, these authors have made it their business to study the media landscape and offer some of the best and brightest perspectives as they peer from the horizon at a new age in journalism and communication.

One of the chapter authors in this collection, veteran new media educator and analyst John V. Pavlik, has written an excellent book about the impact of digital technologies on the field of journalism. Called *Journalism and New Media*, the book combines conceptual and practical information about the changing face of journalism and journalists in the Digital Age. A handful of other books have been written on this subject as well, but *Digital Journalism: Emerging Media and the Changing Horizons of Journalism* is one of the few that showcase knowledge and perspectives from such diverse areas of expertise.

Chapter Summaries

In chapter 1, Kevin Kawamoto asks the question, what is digital journalism? Rather than try to offer a definitive answer, he uses the question as a launching pad for examining a number of different manifestations of the concept and then proposes even more questions that lay the groundwork for further discussion. David Carlson provides the historical context in chapter 2 by taking the reader on a journey through the annals of media technology. He discusses such technologies as teletext and videotex as precursors of some of today's digital interactive devices. Rich Gordon explicates the term *convergence*, a fundamental concept where digital journalism is concerned, in chapter 3. Rather than treat the word as a single concept, he shows that it in fact has many dimensions and applications. In chapter 4 John V. Pavlik explores the tools and the potential of digital journalism in times of crisis. He discusses the role of satellite communications, remote-sensing wireless Internet communications, and mobile information-acquisition devices in helping to get news to people when it is most critical.

Cheryl Diaz Meyer is a senior staff photographer with the *Dallas Morning News* and has taken her digital camera equipment and other accessories into war-torn Afghanistan. She shares not only that experience in chapter 5, but also reflects on the changing role of the photojournalist with the advent of digital photography. In chapter 6, Adam Clayton Powell III shares a speech he gave to media professionals in Canada in which he talks about the easy access to satellite and aerial images heretofore difficult to obtain and how they can be used to supplement and enhance a news story. This access brings with it concerns about privacy, however. Melissa A. Wall is a specialist on the use of digital technologies during social movements. Having closely studied the role of the Internet during the World Trade Organization demonstrations in Seattle in 1999, she discusses "activist digital journalism" in chapter 7.

Paul Taylor was a leader in Washington State's Department of Information Services, an agency that has won numerous awards for its innovative work in the creation of "digital government." Now as a chief strategy officer for the Center for Digital Government, he writes in chapter 8 about the concept of digital government and the "informatics of governing" in helping to build a representative, participatory democracy. In chapter 9, Patricia Radin looks at online health information and the new ways that people are learning about health issues through computer-mediated communication. She offers suggestions to journalists on how to provide "more comprehensive, useful, patient-centered coverage of the larger picture, the issues that ultimately affect everyone in the medical realm." Chapter 10 offers some final thoughts on the concept of digital journalism, which by the end of the book the reader should realize is an enormously broad and multifarious subject.

A Note on References

The Web is increasingly being used as a source of information for conference papers, articles, books, and other academic works. Students, researchers, and scholars should know and follow the expected citation style for the publication or organization to which they are submitting a manuscript. Quoting or referring to Web content for academic work is a relatively new phenomenon, and the citation style pertaining to this dynamic medium is still evolving. In general it is advisable to cite the complete URL, which should take your reader directly to the source being referenced. You should also provide the exact date that you accessed the Web site.

Because this book was conceptualized as a collection of essays—points of view, really—by people who have varying backgrounds in the field of digital journalism, and because the intent of the book was not to publish the results of scholarly research in a narrow sense, the citation requirements used herein are more relaxed than for a scholarly journal. Each author was given some freedom to reference information in a way that best suited the style of his or her chapter—some highly academic, others less formal—reflecting the diversity of expertise brought to this project.

URLs are provided when Web content has been quoted or referred to in the chapters. These Web sites were functioning when the authors accessed them. Since that time, however, the pages may have been relegated to archives and assigned a different URL or removed from the Web altogether. The difficulty in retracing Web content after a certain period of time is part of the nature of this medium. However, even if the

URLs cited here no longer work, if you use specific key words in a search engine like Google, in most cases you will be able to locate additional information about topics covered in this book.

Notes

1. For the full report of this study by the Pew Research Center for the People and the Press, go to http://people-press.org/reports/display.php3?PageID=614. The report, entitled "Public's News Habits Little Changed by September 11," is primarily about use of the Internet for international news but contains useful demographic data.

2. Pew Research, "Public's News Habits."

3. See www.cnn.com/EVENTS/1996/anniversary/how.things.work/index2 .html.

1

Digital Journalism:
Emerging Media and the
Changing Horizons of Journalism

Kevin Kawamoto

A PHOTOJOURNALIST FROM A LARGE AMERICAN NEWSPAPER descends into a war-torn country with her digital camera, laptop computer, and satellite telephone in hand, ready to transmit photographs halfway across the globe for the next day's edition.

A network television reporter in New York City purchases and downloads satellite images from a commercial Web site to enhance his story on deforestation, scheduled to air that evening.

A peace activist takes a digital audio recorder to an antiwar rally and records speeches, group chants, and the personal ruminations of those in attendance. She goes home and immediately uploads her sound files to her personal computer and then transfers them to an independent media Web site that is accessed by tens of thousands of other peace activists from around the country and world.

A group of five journalism students investigates a controversial murder in their city and creates a multimedia Web site to tell the story—and show images of the crime scene—from a number of different angles.

A guy and his friend decide to host a Weblog on which they post news stories that they think are important and interesting from many different sources, assorted bits of daily information about their personal lives, and a way for visitors to the site to leave comments for the Weblog hosts and other visitors to see. Over time the hosts are surprised to find that they have attracted a regular and loyal following. They allow their visitors to post news stories and commentary to the Weblog—virtual strangers sharing their peculiar tastes and observations with each other in a motley community of ad hoc news hounds and pundits.

Arguably, all of these people can be said to be practicing a form of digital journalism. For some, the digital aspect may be a small part of their reportage, which otherwise reflects a more conventional news story. For others, the digital presentation is central to the communication process. It enables a flexibility and creativity that more traditional formats would constrain.

This book is about digital journalism, but it refrains from dogmatic explanations about what, exactly, digital journalism is and isn't. Rather it offers views from a number of people who have given considerable thought to the concept and, in many cases, practiced it.

Not everyone would agree that all of the examples that started off this chapter should reside under the rubric of digital journalism. There are definitions of journalism—digital or otherwise—taught in journalism schools that make presumptions about the practitioners' qualifications and place of employment. This is a debatable subject and thus should be debated, if there is disagreement, rather than dictated. In this age of digital media, where allegedly anyone can be a publisher, the designation of journalist is increasingly being called into question. Who qualifies for press credentials? Who (or what) is the press in the twenty-first century? Should all Web sites that claim to post "news" be designated as news organizations? These are not merely academic ponderings. There have been cases where journalists, or people claiming to be journalists, have been denied press credentials to major news events because the event organizers took a narrow definition of the term *news organization.* Were the event organizers being shortsighted, narrow minded, discriminating, and petty? Or were they forced to establish clear and deliberate criteria for determining what constitutes a bona fide journalist so that every Tom, Dick, and Harriet with a Web site doesn't try to crash exclusive events under the guise of journalistic privilege?

The answers to these questions will have an impact on students who choose to work in nontraditional news settings. Say, for example, that a group of recent college graduates decide to start an online sports news site called SwenStrops.com ("sports news" spelled backwards). They are a team of five, with each person serving multiple functions. They have set up a Web site and aim their content at college sports fans. Every employee does some writing for the site, but he or she might also perform design, programming, moderating (e.g., electronic discussion groups), marketing, event planning, or other responsibilities. They have about three thousand unique visitors to their site per month, which is a drop in the bucket compared to online sports giant ESPN.com. Is this really a news organization? Are the employees journalists? Shouldn't the editorial and marketing functions of the organization be separate? Should the staff members be issued press credentials so they can be admitted to sporting events free of charge? Why or why not?

As you consider this hypothetical example, keep in mind that Bloomberg Business News (BBN), which today is considered a leading financial news organization, started off relatively unknown in the world of financial reporting. Over time it gained exposure, credibility, and acceptance. How do we know SwenStrops.com isn't another BBN in infancy? In these information-saturated times, how do we decide who and what are legitimate sources of news and information, and who and what are not? Are these questions themselves problematic? Do they offend the principles of a free press? Or are they necessary to protect the integrity of journalism as an independent social force? What other questions should students of journalism in the twenty-first century be asking?

If this book is successful, it will get readers to think about the challenges and opportunities of working in the field of digital journalism. There is much practical information to be learned here, as well as much food for thought. Use it as a means to stimulate your thinking about the field and to question old and new assumptions about what the practice and theory of journalism should be.

What Is Digital Journalism?

Digital journalism is a difficult concept to define precisely because it can mean different things to different people. In fact, the term is a juxtaposition of old and new concepts. Journalism as we know it in the West goes back at least two millennia with the appearance of the *Acta Diurna* ("Daily Events"), official texts of general interest handwritten and publicly posted in ancient Rome during the time of Julius Caesar. Conversely, digitization, the process of converting information to a computer-readable format, was born in the age of the electronic computer during the twentieth century and requires advanced technology for distribution and display. Taken together, these two words suggest an old practice in a new context—a synthesis of tradition and innovation.

At the risk of coming across as too narrow to some and too broad to others, this book proposes one definition of digital journalism: the use of digital technologies to research, produce, and deliver (or make accessible) news and information to an increasingly computer-literate audience. This definition captures the historical and important function of journalism in a democracy, which is to inform and enlighten the public, and also acknowledges the evolving tools of the trade and literacy skills of the audience. These tools can impact how journalists and their news organizations research news events, move information from one place to another, construct and organize stories, develop interfaces, and otherwise engage their audiences. Aside from the electronic, digital nature of the information processing, the following characteristics are typical of digital journalism:

- *Hypertextuality:* the linking and "layering" of digital information through a nonlinear hierarchical structure
- *Interactivity:* the process of engaging active human or machine participation in the process of information seeking and information sharing
- *Nonlinearity:* a flexible ordering system of information that does not necessarily adhere to traditional, chronological, or conventionally logical patterns of storytelling
- *Multimedia:* the use of more than one type of media in a single product
- *Convergence:* the melding or blurring of historically discrete technologies and services
- *Customization and personalization:* the ability to shape the nature of the content and service to individual needs and desires

Digital journalism includes all of these things and more. The definition of digital journalism is a moving target. As technology changes and, more importantly, as the institution and concept of journalism change, the definition of digital journalism will also change. Do you agree with this statement? Does your definition of digital journalism resonate with the one provided in this paragraph? If not, how would you change it?

Although it is tempting to treat the Internet as synonymous with digital media because it is currently the dominant delivery system of digital content, it is really only a subset of digital media. Digital media encompass other technologies such as digital television, personal digital assistants (PDAs), Web-enabled phones, satellite-based technologies, portable computer "tablets," and a host of other devices that may not even be on the market yet. Because the Internet and the Web are the most visible manifestations of digital media, they will dominate much of the discussion in this book.

The recipients of this news and information tend to be more than just passive receivers of content. Internet users are not referred to as "couch potatoes." In fact, to describe them as "recipients" may be misleading: They are more appropriately described as "information seekers," comfortable with search and navigation strategies on the Internet and with other aspects of computer-mediated communication (CMC). However, traditional media users who are not computer literate can also benefit from digital journalism in cases, for example, where a print or broadcast news story was the product of a computer-assisted research project or a digitally produced graph or illustration. Digital journalism, more loosely defined, could simply involve the use of digital technologies to produce content for a general and traditional audience. But the more computer literate information seekers are, the better able they will be able to exploit the resources produced through all forms of digital journalism.

Coming of Age

Those who regularly use the Web to get news and information sometimes find it hard to believe that online newspapers were rare in the early 1990s. Only a small number[1] could be accessed via a networked computer—some through electronic bulletin board systems (BBSs), others through fledgling commercial Internet service providers (ISPs). The *Columbus Dispatch* in Ohio began offering an online edition through CompuServe in 1980, but the ISP was relatively small at the time, serving fewer than four thousand members.[2] The number of people using the Internet was growing, but had nowhere near approached the numbers that were to materialize from the mid-1990s onward after the rapid adoption of Web browsers by a burgeoning population of Internet users. After that, the *San Jose Mercury News* was one of the first newspapers to have a major presence on the Web—namely, the *Mercury Center*—first through America Online (AOL) in 1993, and then directly on the Web. Others followed suit in fairly rapid succession after 1995.

Today the more than three thousand U.S. newspaper Web sites (and thousands more abroad) are pretty much taken for granted by inveterate Web surfers. Rarely do people marvel at the mere presence of a newspaper Web site anymore; rather, it is expected that a newspaper should have an online corollary, and most of them, even relatively small ones, do. According to *Editor and Publisher* (E&P), during the month of June 2002, half of the top twenty news Web sites were affiliated with newspapers.[3] Many magazines, radio stations, and television stations also have their own Web sites, which, like newspaper Web sites, contain repurposed content from their traditional print or broadcast division, as well as new content unique to the online medium. Web sites are no longer crude duplications of their parent company's traditional news product, as they once were. The term *shovelware* referred to the software that electronically shunted content from traditional newsrooms to the new media division's computers for eventual posting on a BBS or Web site, often without much attention to design elements.

In the early days, online media content tended to be redundant (vis-à-vis their traditional counterparts) and did not fully tap into the potential that a digital medium offered. As online news media matured, they also came of age, so to speak, acquiring their own identities, styles, looks, and relationships with audiences, thereby distinguishing themselves from their nononline news counterparts, and yet complementing them at the same time. For this coming of age to occur, those who manage and staff online media divisions at traditional news organizations had to ask themselves questions like, What is the value-added of digital media? What can we do differently or better with digital media technologies that cannot be done in traditional news formats? The answer to these important questions helped the online news media discover their own assets and identities as distinct siblings to their traditional brethren.

In the twenty-first century it is clear that digital journalism is entering the mainstream. As mentioned earlier, it is being incorporated into journalism curricula as those who oversee developments in academia realize it is not only pedagogically responsible to do so—since the field of journalism is becoming more digitally oriented—but students also seem genuinely to enjoy these classes. It was not always the case that a student could take a photograph and immediately publish it for the world to see on a Web site. So-called wet processing, the conventional way of processing photographic film using liquid chemicals in the development process, is time consuming and relatively messy. Now digital cameras not only allow photos to be downloaded quickly onto a computer as a digital file, but if that file needs to be transmitted hundreds or even thousands of miles away, it can be sent instantly through a telephone line or a satellite uplink-and-downlink procedure. Photojournalists in rugged conflict zones where reliable communication and transportation services do not exist can now "rig up" their own long-distance personal-communication system using a digital camera, battery-operated laptop computer, and satellite telephone or videophone. These journalists would not necessarily describe themselves as digital journalists—they are simply journalists, without fancy adjectives—but they are doing their jobs with the help of digital technologies and, hence, engaging in a form of digital journalism.

There are also online journalists who work primarily or exclusively for an online news organization unaffiliated with a traditional news company. In other words, their digital news divisions did not sprout out of an existing newspaper, magazine, or television show, but originated as an online service. Slate, TheStreet, CNET News, Salon, WebMD, and others are examples of these. Not-for-profit media foundations like The Freedom Forum and the Poynter Institute have journalism news Web sites, as do thousands of other organizations with their own particular content focus (e.g., human rights, labor, environment, health, technology). The field of digital journalism is organizationally and topically diverse.

There is a growing number of resources devoted to digital journalism. For example, the Annenberg School for Communication at the University of Southern California has published the *Online Journalism Review* (OJR) since March 1, 1998. The online journal's purpose, as explained on its Web site, is "to be useful to journalists and anyone interested in where journalism is going in cyberspace."[4] Other journalism-related organizations such as the Poynter Institute, The Freedom Forum, the American Society of Newspaper Editors (ASNE), the American Press Institute (API), the National Institute for Computer-Assisted Reporting (NICAR), the Pew Center for Civic Journalism, and others have devoted significant resources to helping traditional journalists and news organizations understand and use new media technologies. Magazines and journals like *Broadcasting and Cable*, *Columbia Journalism Review*, *Editor and Publisher*, *Journal of Broadcasting and Electronic Media*, *Nieman Reports*, and *Quill* publish either occasional articles or regular columns (or both) that discuss developments in the digital media environment.

In associating with the venerable Columbia University Graduate School of Journalism (J-school), the Online News Association (ONA), founded in 1999, has played an important role in bringing legitimacy and credibility to the field of digital journalism. Since May 2000, the ONA and Columbia's J-school have sponsored the Online Journalism Awards. As explained on the organization's Web site, "The contest honors excellence in Internet journalism and is open to all English-language Web sites around the world."[5] In 2000, awards were given in eight categories: General Excellence in Online Journalism, Breaking News, Enterprise Journalism, Service Journalism, Feature Journalism, Creative Use of the Medium, Innovative Presentation of Information, and Commentary.

The New Media Federation of the Newspaper Association of America (NAA) has sponsored the Digital Edge Awards since 1996. This annual competition has recognized outstanding achievements by online newspapers with categories such as Best Local Online Service, Best News Presentation, Most Innovative Use of Digital Media: News Event Coverage, Most Innovative Use of Digital Media: Features/Enterprise, Best Vertical Site, Public Service, Best Classified Site, Best Advertising Program, Pioneer Award. (Award categories have evolved over the years, possibly resulting in different categories than those listed here.) Other awards are given for distinctive Web sites, some of which include digital journalism projects.

Modeled after more mainstream award events, the Webby Awards are given by the International Academy of Digital Arts and Sciences each year. There is a fee for entering a site in this competition, and judges base their decision on six criteria: content, structure and navigation, visual design, functionality, interactivity, and overall experience. A refreshing novelty of the awards ceremony is that winners are allowed only five words in making their acceptance speech. The 2002 Webby Award Winner for News was the online version of *BBC News*. The EPpy Awards are given each year at E&P's Annual Interactive Newspapers Conference and Trade Show. The awards recognize and honor "outstanding achievements in new media by newspapers," according to E&P's Web site, as well as presenting an award for outstanding individual achievement. Regional chapters of the long-established Society for Professional Journalists also give awards for online news. These award events and others reflect an industry that is trying to define and reward excellence among its members while establishing a collective set of standards by which to judge "quality" news Web sites. This process may be viewed as elitist and narrow—resulting in the canonization of cookie-cutter design and content concepts—but it also reveals the development of a mainstream online news culture, which by definition would entail the establishment of in-group norms and values. Like in the real world, however, mainstream cultures compete with countercultures and subcultures and their opposing or conflicting ideas. This phenomenon would be particularly applicable on the Internet, where the diversity of organizations, individuals, and perspectives is a defining characteristic of the medium. Digital media—the design, content, interface, and so forth—are particularly vulnerable to "renovation." Unlike traditional media, whose mastheads and fonts and layout design are fixed to establish a visual identity over the long run, the culture of digital media is much more flexible in this regard. The look of Web sites often changes (sometimes at the whim of new designers who have joined the company), and this change is not considered a sign of confusion or indecision. Perhaps no other news medium is more malleable than the news Web sites. Web designers and decision makers are constantly trying to build a better mousetrap, so to speak, and are influenced by what other designers and decision makers are doing. The notion of "copying" does not have quite the pejorative connotation that it might in some circles. This does not have anything to do with plagiarism or copyright violation, but rather with Webmasters looking at what other Webmasters are doing and getting ideas for their own Web sites—a kind of "hey, that's neat" and "let me try that" attitude.

As digital journalism has come of age, many consumers of news and information have also sharpened their multimedia proficiencies. They have access to digital video and photo galleries, multimedia slide shows with photos and audio clips, comprehensive in-depth special reports that have much more information than could possibly be contained in a single print or broadcast report, e-mailed news alerts about breaking news, and customized or personalized news services. The *Seattle Post-Intelligencer* (*P-I*), for example, allows Web site visitors to register for a free service called PImail, which "delivers the latest news of interest to you straight to your inbox." There are dozens of categories (or what the *P-I* calls "channels") to choose from, such as local news, food and dining, Microsoft, and Huskies (football). PItoGo allows people to download news articles directly to a PDA,[6] a handheld electronic device. PImobile delivers "the latest seattlepi.com headlines, weather forecasts and traffic incidents to your Web-enabled cellular phone." And PIdesktop offers "a continuous stream of news information from your favorite sources direct to your desktop in a compact, customizable ticker toolbar. Desktop News is efficient, requiring little memory or bandwidth and won't take over your desktop like other less polite applications."[7] Obviously, digital journalism is not confined to the Web. In addition to handheld devices, news and information in digital form will likely end up on a variety of devices from computer–television hybrids to satellite radios. Right now, as mentioned earlier in this chapter, the Web is the dominant delivery system, but that may not always be the case. Perhaps something will take its place in the future—something wireless and portable, but capable of handling large amounts of information quickly and seamlessly. Digital media analysts are always trying to predict what device will be the "next big thing."

Cooperation and Competition

When traditional media organizations first began creating online news divisions, there was not always a clear understanding of what the relationship between the traditional and digital media divisions should be. Understandably, the digital media division was often perceived as being a competitor and threat. Online newspapers could easily "scoop" their print counterparts by posting breaking news on the Web site far ahead of newspaper delivery of the same news to homes and newsstands. Also, a nagging question perturbed those who were uncomfortable with the presence of online newspapers: If people start reading their news online, why should they continue buying or subscribing to the newspaper? Maybe the online version of the product would replace the old-fashioned newspaper and, worse, supplant traditional journalists with a bunch of young and cocky Web heads?

Online staff members were frequently segregated from the traditional journalists, sometimes for simple lack of physical space, which only widened whatever ideological rifts already existed. Online staffers sometimes looked down at their traditional print counterparts as well, seeing them as defending anachronistic practices that were inhibiting media innovation. Over time, however, these fissures seemed to mend as it became apparent that the relationship between traditional and digital media existing within the same news organization was obviously symbiotic. Digital media require content, and journalists produce content (although many dislike being called "content producers"). Traditional newspapers, for their part, were getting signals that their industry was in decline, especially with the younger generation of information seekers, and that they would continue to suffer unless they embraced a number of technological, economic, and organizational innovations. Newspapers needed an image overhaul as young news consumers flocked to a television diet of colorful visuals and brisk storytelling. Digital media was hip, savvy, and cool, and fortunately, it could also be substantive so as not to entice audiences with bells and whistles alone.

The traditional media and digital media are increasingly working in tandem these days, cross-fertilizing (or cross-promoting) each other by driving audience members back and forth between them. The broadcast news media are using the Web in creative and complementary ways. PBS, for example, has long directed viewers to visit its Web site for more information about things that could not fit into the scheduled time slot of the televised program. Other stations also commonly use this practice to give their viewers more information about news topics that may have been considered extraneous for broadcast. This is a good way to develop audience relations and use content that did not make it on air, but has value to those viewers who are interested in learning more about a story. It is not uncommon to hear news anchors and television magazine hosts encouraging viewers to visit the program's Web site, where they can find more information as well as participate in electronic discussions, ask questions of experts, send e-mail to the reporters, learn about special effects used in the program, and so forth. MSNBC has links to the "Today Show," "Nightly News," "Dateline NBC," and "MSNBC TV" home pages. ABCNEWS.com has links to "Good Morning America," "World News Tonight," "20/20," "Primetime," "Nightline," "UpClose," "This Week," and "ESPN Sports" home pages. Cbsnews.com has links to "The Early Show," "CBS Evening News," "48 Hours," "60 Minutes," and "60 Minutes II" home pages. Almost all television Web sites have detailed programming information, video clips of upcoming or past shows, and sometimes live coverage of breaking news. The cross-promotion between the traditional and digital media components of a single media organization is a constructive way that these two divisions can coexist, each helping to drive audience traffic to the other side. As digital media continue to grow as venues for the news and information needs of the public, the traditional news media and their digital media offshoots will need to see themselves as essential to each other's long-term survival.

One model of media interconnectivity is the portal Web site, where the Web user can access many different Web sites through one clearinghouse site. Bayarea.com, for example, offers links to the online services of the San Jose *Mercury News*, the *Contra Costa Times*, the *Monterrey Herald*, the *Viet Mercury* (Vietnamese-language news), and the *Nuevo Mundo* (Spanish-language news). Sometimes these clearinghouses are the result of chain-driven initiatives, where the large media company that owns newspapers in many different markets tries to link them together through a portal. Sometimes they are the result of different types of media in a local or regional community getting together so that Web visitors can access newspaper, TV, and radio Web sites all at the same Web address.

Cyberforums

Electronic message boards allow viewers to comment on or criticize how a subject was covered on television or in a newspaper or magazine. To discourage flaming wars and specious personal attacks, many message boards require that users register first and abide by certain ground rules or risk having their comments removed or not posted at all. Without these safeguards, and sometimes even with them, discussions on these boards can become unruly and disrespectful.

Nevertheless, electronic discussions can be informative and helpful under the right circumstances. Historically, news stories have always helped generate conversation among people who are interested in the subject matter. People talk about currents events in their own family circles, among friends, around the office water fountain, on the bus with strangers, in classrooms, at the town hall, and elsewhere. Ideally, these discussions help people formulate and revise their own perspectives by creating a forum to articulate their ideas, listen to what others have to say, and respond to different points of view. Electronic discussion groups expand that conversational space to an electronic, conceptual dimension. Whether in the real world or in cyberspace, the more controversial the topic, the more spirited the discussion tends to be. It is clear to anyone who has actually participated in electronic discussions on the Internet that some discussion groups are more useful than others. Some message boards, especially when they are not moderated and have no guidelines for civil discourse, can disintegrate into a space for vituperative name-calling and unabashed displays of ignorance. But they are certainly not all that way. Some are meaningful virtual communities, where people who participate in discussions do disagree, but do so respectfully, and where others offer support, advice, information, a "listening" ear and virtual companionship. Relationships that develop in these spaces do at times extend into the physical world. More often these discussion groups are just a place to "chat," to give an opinion, or to share a story or two.

In July 2002, ABC's national television news magazine, "20/20," reported on what seems to have become a national obsession: obesity and weight loss. The program raised questions about the wisdom of following a low-fat, high-carbohydrate diet, which most of the medical community and U.S. government have been endorsing for years. It suggested that the Department of Agriculture's food pyramid, which encourages people to consume little or no fat and relatively larger portions of carbohydrates (compared with the other food categories of proteins, fruit, and vegetables), might be mistaken and actually lead to weight gain rather than reduction.

On the network's Web site, there was a link to the "20/20" home page and hypertext links to other articles about various diets such as the Atkins and Zone diets, both of which diverge from the conventional wisdom. A transcript from a live chat with ABC's Dr. Tim Johnson was also available, as were transcripts of chats with other health experts. There were options available on the Web site to e-mail the story to a friend, to check to see what story on the site had been e-mailed the most frequently, and to chat with other viewers on an electronic message board. On this board were comments by people who advocated the high-protein diet and those who criticized it, and others whose views fell somewhere in between. One person claimed to have lost sixty-five pounds on such a diet and wrote, "at last I know I have found the answer." Others offered alternative diets that worked for them, and still others outright rejected the high-protein diets, some claiming that such a diet would lead to long-term medical problems. Ultimately, the Web site gave visitors access to a lot of information about diet and nutrition—including links to expert opinions and first-hand claims by non-experts—and then left it up to the individual to decide what theories and approaches were most plausible.

The ABCNEWS.com Web site is just one example of a traditional broadcast company using the Web to supplement its traditional fare and engage its audience members in ways not possible or feasible before the advent of the Web, e-mail, electronic message boards, and other digital media mechanisms. Most other traditional news organizations have similar digital media services. MSNBC allows its Web users to rate articles and vote for their favorite photos of the week. After voting, visitors can see the aggregate top-rated articles and favorite photos and compare those results with their personal preferences. This process is not a scientific method of collecting group responses, but it is a fun way of building interaction into a news Web site. CNN Interactive provides links to program transcripts. Other interactive devices include flash polls (for quick public-opinion surveys, which are, again, unscientific), search engines, games, puzzles, stock quote search windows, customized e-mail notifications, video news clips, audio files, e-commerce, local weather reports, and much more. Although these companies are national and international in scope, they provide access to local news sources, usually by partnering with local television stations.

Of course, electronic discussion groups do not have to be sponsored by established news organizations. Free-standing electronic discussion groups (e.g., so-called newsgroups) long preceded those hosted on news organizations' Web sites. They exist for a variety of reasons—information sharing, companionship, support, romance, advice, commerce, criticism, illicit exchange of digital content, political activism, and so forth. (These are not exclusive or exhaustive categories.) The content circulated in these groups often provides alternative or supplemental sources of news and information, albeit with differing degrees of credibility, than that collected and disseminated by the mainstream media. They obviously feel the communication and information needs of a segment of the Internet-using population.

Small Media

The large commercial news media are not the only ones taking advantage of digital media technologies. The Web is particularly beneficial for non-profit and grassroots media that do not have large financial reserves to fund a global media effort. An organization like the Independent Media Center (IMC) has a Web site that offers an alternative to corporate-owned news media. Established in 1999 to cover the World Trade Organization meetings and demonstrations in Seattle, it now calls itself "a network of collectively run media outlets for the creation of radical, accurate, and passionate tellings of the truth."[8] The IMC serves as an information clearinghouse for journalists and can make reports, photos, audio and video files, and documentaries available through its Web site using Apple QuickTime and RealNetworks technologies. "Reporters" anywhere in the world can take photos or videos of a demonstration, write about it, and e-mail the full package for quick posting on a Web site. This is why digital media are often regarded as a great equalizer, allowing "small media," which have far fewer resources than their media-conglomerate counterparts, to have Web sites that are as technically and aesthetically sophisticated. Volunteers and low-paid staff have always been important in keeping the small media alive.

Another curious digital journalism phenomenon has been the emergence of Weblogs, or "blogs," which are Web sites run mostly by individuals (as opposed to professional news organizations) that are updated regularly using content management software, which makes adding fresh information and links to other Web sites relatively quick and easy. An example of a Weblog, of which there are reportedly hundreds of thousands, is the Scripting News site at www.scripting.com. A blog is difficult to describe because they are not all the same. It resembles an online journal or diary with hypertext links to topics that the blog's owner finds interesting or thinks others will find interesting. Typically, there is a chronological column of log entries with posts that occur daily, more than once a day, or less than once a day. For example, if someone's blog attracts visitors who are interested in the First Amendment, that person might write a comment about a specific Supreme Court decision that was just handed down a couple of hours before and link to an article that discusses the decision. Or the blog could be much more personal and seemingly trivial. A college student has a hot date for the evening and writes in his blog that he is going to a nightclub and does not expect to be home until late. Why someone would announce such a thing on the Web is not always evident, but apparently there are people who find such information worth reading (e.g., family or friends of the student, perhaps). One blog, www.fark.com, contains a daily collection of many different articles from a diverse range of media sources. To the uninitiated, it is difficult to explain what a blog is or whether it is of any news value. One simply has to visit a few to get the idea.

Whether amateur reporting, commentary, and blogs qualify as digital journalism is open to debate. Some would argue that those who engage in digital journalism should be professionals, trained as journalists and working for a "legitimate" news organization. But what is a legitimate news organization? The question of who qualifies for press credentials these days is becoming a thornier one in a digital media environment. A *New York Times* reporter would not have trouble getting press credentials to cover an exclusive political event, but would an IMC reporter be issued such a permit? It is really up to the credentialing organization. In this world where just about anyone can be a (Web) publisher, the issue is a tricky one. What constitutes a bona fide journalist? A number of instances have arisen in recent years where Internet-only journalists (with no affiliation to a traditional news organization) have been denied press credentials to cover a particular event. A controversy was sparked in Germany in February 2000 when many online journalists were not issued press credentials to cover, ironically, a massive international computer and Internet trade fair. A representative of the trade show explained why: "The problem is [that] on the Internet, everyone can do his own homepage and put content on there and tell us that he is an online journalist."[9] The International Olympic Committee (IOC) decided not to credential Internet reporters for the Olympic games in Australia in 2000, even when the journalists were affiliated with traditional news organizations. The IOC did credential reporters from around the world who were affiliated with traditional print and broadcast companies. In addition to the difficulty of having to decide who is a legitimate journalist and who is not, the IOC thought that having Internet journalists there might compromise their agreements with television stations that paid for broadcast rights.[10]

In June 2000, *USA Today* reported on the problem confronting many event organizers regarding who should and should not be given press credentials. A number of online journalists—that is, those who wrote for Web publications—were quoted in the article as having been denied access to news events or, in the case of a reporter writing for a financial news Web site called TheStreet.com, being "escorted out" of a high-profile investment symposium.[11] Even veteran journalists with years' worth of distinguished service with the traditional media who had made the move to an online publication complained about being denied press credentials. What are the criteria for establishing whether a person qualifies as a journalist or not? There are no universal codes. Event organizers create the criteria, which can range from highly restrictive (so as to eliminate all online journalists) to widely accessible (with few obstacles to gaining press privileges). Here is an example of one organization's criteria at one time. The event has to do with Internet developments, but its press credentials are relatively restrictive. To qualify, one must be affiliated with a nationally or regionally recognized media outlet and hold an editorial title or hold a position as an industry analyst. The list of criteria[12] (adapted from the organization's press release) continues. To obtain your press credentials at the show, you must present the following:

- Your media outlet's masthead, stationery, or business card, with your name and editorial title

- A bylined article written by you and published within the last six months
- A letter from an appropriate publication and signed by an editor or publisher of that publication assigning you to cover Streaming Media West 2002 or Internet World Spring
- For freelance journalists, photographers, or videographers, a letter from an appropriate, recognized media outlet on its stationery stating that you have been retained to cover Streaming Media West 2002 or Internet World Spring.

The following people do not qualify for press credentials:

- A publisher
- A sales rep
- A freelance photographer not affiliated with a recognized, appropriate media outlet
- Any other noneditorial personnel
- A representative of a personal Web site

It wouldn't be difficult to exclude journalists who exclusively write for independent Web publications from this event. Many of the terms are open to interpretation. Even though it is theoretically true that with the Web anyone can be a publisher or journalist, a large amount of gate-keeping still goes on to separate the established media from the more avant-garde, independent media, or just to keep the dot-com side of traditional media organizations at bay. As new boundaries are drawn, it is not always clear who gets to play in the sandbox and who has to stand outside and watch. The rules have been in a state of flux. Policies about press inclusion and exclusion one year might be modified, abandoned, or reversed the next. This is a reflection of how difficult it is to define the "real media" in a digital age.

Community Web Portals

At one time, new dot-coms threatened to create Web services on the Internet that would provide information to audiences traditionally served by local newspapers. For starters, users could get restaurant and movie reviews, information about the local arts and entertainment scene, and the prices and availability status of local accommodations. The chief revenue stream for a newspaper, classified ads, would probably not be far behind. A number of companies without a traditional news background were penetrating local markets and making information available that people used to get by picking up a local newspaper. There was a possibility that these dot-coms would steal readers and revenues away from local newspapers, competing for local advertising and creating dynamic "community spaces" online that newspapers were not providing. There was no telling where this could lead. What would stop a dot-com with no ties to a local community from stepping in and setting itself up as the premiere online hub for local news, advertising, chat, reviews, jobs, and so on?

In response, some local newspaper Web sites began creating their own community Web sites, serving as a kind of portal for local arts and entertainment information, community and neighborhood news and organizations, and links to jobs and other opportunities. The Tribune Company, which publishes the *Chicago Tribune*, created its own city guide Web site, called Digital City. Knight Ridder created RealCities.com. Myway.com, Zip 2, and other companies were among the first movers in helping newspapers develop community-based Web sites affiliated with their existing online newspapers, but more focused on communities and neighborhoods. An example of such a site is SouthSound.com, sponsored by the *Tacoma News Tribune* in Washington State. This site was designed to "connect communities in the South Puget Sound" region and offered space online for community organizations to put up their own Web sites to let others in the community know what organizations exist and what events were coming up. MySanAntonio.com contained content from a local San Antonio, Texas, newspaper and television station, but also provided information pertinent to community and neighborhood organizations such as access to a San Antonio crime database (searchable by neighborhood) and, like the SouthSound.com, links to community organizations and articles about neighborhood news. Similar community sites emerged elsewhere in the country as distinct entities from the local newspaper Web site, although owned by the same media company. These sites have met with mixed success. Myway.com acquired Zip 2 and eventually restructured, but sites like SouthSound.com continue to operate (as of this writing).

Others have discontinued their community sites. The general sentiment is that community Web portals have not lived up to their expectations, and it is not clear whether and how these sites will evolve in the future.

Computer-Assisted Research

The impact on journalism of computer-assisted research (CAR), also known as computer-assisted reporting, has been significant. Succinctly put, CAR has not only increased the power of journalists to do their jobs better with innovative analytical and technological tools, but it has also allowed nonprofessionals—such as amateur news hounds and the general public—to access and analyze information on their own without relying on an interpreter or intermediary to get between them and a particular set of data. Hence, CAR is both a tool for news professionals and for the proverbial person in the street (who has the proper training). These days, many journalism schools offer at least a basic CAR course for their students to take, but the content of such courses may vary widely.

CAR, loosely defined, could be as simple as learning how to use search engines to find out information about a particular topic on the Internet. Or it could involve electronically sifting through hundreds of thousands of computer files to try and match up, say, records of people convicted of felonies in a particular city and records of people licensed to run childcare businesses in that same city. A potential finding from such an analysis would be the government's lax regulation of childcare providers. Done manually, without the use of computers, such an analysis would be unwieldy and time consuming. Using relational databases—where a software program finds relationships between records in one database and records in another—the process may take only a matter of minutes. On second thought, while technically true, the previous statement is misleading. A computer may literally take minutes to do the actual number crunching, but it may take days, if not weeks or months, for the researcher to acquire the necessary records and prepare them for this relatively quick analysis. Nevertheless, analyzing hundreds of thousands of records is generally quicker and more accurate with a computer than with humans who are prone to error due to fatigue, carelessness, bias, or other problems.

Not much needs to be written about the importance of CAR here because there are a handful of excellent books and Web sites that discuss this critical competency area for journalists and journalism students. Professor Philip Meyer of the University of North Carolina, Chapel-Hill, might be called the grandfather of CAR because of his pioneering work in this area back in the late 1960s. His book, *Precision Journalism: A Reporter's Introduction to Social Science Methods*, was first published in 1973 and has been revised several times since. After receiving specialized training, Meyer used social science methods to research the 1967 Detroit riot for the *Detroit Free Press*. Since then he has been a passionate advocate for teaching journalists about social science methods, including computer-assisted analysis and reporting of raw data.

CAR has come a long way since the 1970s, when much electronic data was stored on mainframe computers and computer-based analysis was cumbersome and relatively unfriendly. Today CAR is easier to conduct on personal computers using the Web or CD-ROM databases, Windows-based statistical analysis software (e.g., SPSS), spreadsheets (e.g., Excel), relational databases (e.g., Access), hardware accessories, and software applications. These new tools may make CAR easier, but telling good, accurate, and balanced stories with analyzed data is still essential. A solid understanding of social science methods helps the journalist understand and correctly analyze and interpret raw data. CAR stories have included analysis of databases of all kinds, including crime, campaign spending, population, housing loans, health, and incarceration. For examples of news stories using CAR, visit the National Institute for Computer-Assisted Reporting Web site (www.nicar.org), and the Investigative Reporters and Editors Web site (www.ire.org). The Poynter Institute's Web site also features resources and information and training opportunities for journalists and journalism educators interested in CAR.

In addition to Meyer's book, which has a 2002 edition, books on CAR include Bruce Garrison's *Computer-Assisted Reporting* (1998), Brant Houston's *Computer-Assisted Reporting: A Practical Guide* (1998), Lisa Miller's *Power Journalism: Computer-Assisted Reporting* (1997), Margaret DeFleur's *Computer-Assisted Investigative Reporting: Development and Methodology* (1997), and Nora Paul's *When Nerds and Words Collide: Reflections on the Development of Computer-Assisted*

Reporting (1999). These books and Web sites give a sense of the tremendous public service CAR can provide, and it is no coincidence that most of these works appeared after the popular adoption of the Web, where information can now be accessed directly from government, university, corporate, nonprofit, and even private computer servers.

CAR is integrally tied to digital journalism because of the use of technology and journalism to produce stories that have a public interest. But CAR can result in a traditional news product, one that does not have to appear in a digital medium. If it does appear in a digital medium, the possibilities for presentation and for interactivity are exciting. Web users can read a CAR story about crime statistics, for example, and then search a Web-based database for information about crime in their own neighborhoods or another neighborhood of interest. Another possibility: A reporter could write a story based on a search in the opensecrets.org Web site, a user-friendly database sponsored by the Center for Responsive Politics that provides data about campaign contributions and political candidates. Then Web users could be encouraged to use the Web-based database and do their own searches, maybe about politicians in their own neck of the woods. Other databases available on nonprofit, government, or educational Web sites could allow for similar kinds of interactive sleuthing, but of course, the results are only as reliable as the databases and research process used to get those results.

Investigative journalism using CAR has won some of journalism's highest awards. More importantly, reporters have exposed corruption, waste, unsafe practices, fraud, discrimination, environmental hazards, and other social problems by tackling hitherto monumental tasks of data analysis. Typical of CAR is the close examination of records of people hired into positions of trust. Such investigations have resulted in exposés of employees who have long criminal histories that were not reported on job application forms, thereby implicating the efficacy of existing screening procedures. Evidence for racial profiling also has been gathered through CAR. The list goes on.

CAR is one of many aspects of digital journalism. As mentioned earlier, it is a powerful analytical and technological tool, and one that is accessible to both professional journalists and nonprofessionals who have the skills to use the tools. It is a good example of how journalism and technology can well serve the public interest.

Funding Models

In traditional news organizations, there is a metaphorical firewall between the advertising and editorial functions of a news medium. Although this book is not about the business of digital journalism, a brief discussion of the economics of digital media is appropriate. Digital news divisions (including staff, technology, content, services, and marketing) are expensive to support and, with few exceptions, have found it difficult to generate adequate revenues to justify their existence and growth. Media organizations have tried many different funding models, from banner ads to paid subscriptions, with disappointing results. Banner ads are notoriously unsuccessful, and people do not usually pay for news unless the timeliness of the news (financial information, for example) is of high value in itself. As of July 2002, the *Wall Street Journal's* online service cost $79 per year or $39 for those who already subscribed to the print edition of the newspaper. That funding model works because many people, such as investors and entrepreneurs, are willing to pay for trusted, up-to-date financial information. CNN Interactive, after years of providing free video clips of news on its Web site, began charging for their video in 2002. "CNN Video is now available only with a subscription," a pop-up window explained when users clicked on the video hypertext. In July 2002, cost for the video option ranged between approximately $5 and $40 a month, depending on the video package desired. At the same time, MSNBC still offered its video services without charge.

Many online newspapers offer all of their content without charge. Others provide some of their news for free and charge for premium service. More organizations are charging for archived articles, although their current news is free. Other nonbanner advertising can be found on most news Web sites, but few claim to be big winners in the online advertising game. Pop-up ads are annoying and have the potential to backfire as people intentionally avoid sites that irritate them with tricky advertising ploys. More ideas are needed. In late 2000 and early 2001, one media organization after another began announcing heavy reductions in staff and even closures of their new media ventures. But not everyone was overly concerned. *Business Week* magazine was more circumspect. In a February 2001 news analysis, it wrote:

> More than a retreat, in fact, the cutbacks are the inevitable acknowledgement that Web businesses can't live a financial fairytale forever. Record print-publication earnings over the past two years afforded these media behemoths the luxury of indulging in lavish experiments without incurring shareholders' wrath. Now, the bash is over, and they're consolidating operations and cutting costs.[13]

The bottom line is that the financial outlook for digital media organizations is not rosy. For years, the digital media divisions have been promising profitability on the horizon, but a series of economic downturns in the nation, combined with many poor business plans, have rendered those promises undeliverable. The solution was not massive abandonment of digital media divisions but a scaling back and rethinking of how to proceed. The financial viability of digital media remains a lingering concern with no easy answers. Many factors have to be taken into consideration, some of which cannot be foreseen or predicted, like the economic fallout from acts of terrorism or the lack of investor and consumer confidence as the result of highly publicized corporate scandals. The competition for ad dollars during times of economic belt-tightening is also a key factor.

An entire book can be written about the complexities of digital media economics. The same conundrums that surround any other business plague digital media businesses as well in volatile economic times. News organizations are increasingly run like large corporations, with attention being paid to quarterly profit earnings, market share, and ratings. Digital media are particular sensitive to fluctuations in earnings because they are relatively new entities; in many cases, they have not yet established their centrality to their parent company's operations and, thus, are vulnerable to cuts and reductions. (For a broad overview of the digital media environment, see Kawamoto's *Media and Society in the Digital Age*, 2003.)

But digital journalism will not disappear because, ultimately, it is becoming an essential component in the contemporary news environment. In a relatively short period beginning in the mid-1990s, millions of people have adopted the Internet as their primary or secondary source of news and information. The business side of digital news organizations needs to figure out how best to convert those numbers into dollars. Even if this is not immediately possible, digital news sites are still valuable insofar as they increase brand name visibility, which benefits the traditional news product, and provide interactive, engaging content that can build more positive and loyal audience relations. For these reasons and more, digital news media will thrive, despite current economic anomalies.

The Basics and Some Questions

For all the talk about the "digital" in digital journalism, there is the potential to lose sight of the basics. Good journalism, regardless of the medium, still entails telling stories that are well researched, engaging, based on facts, accurate, fair, balanced, carefully proofread, properly contextualized, ethical, and readable. Humans are storytelling creatures, and writing for the news media is an act of storytelling: Something happened that is important for other people to know about. There are different reasons for telling a story. In news, it is usually to inform and explain, but it can also be to entertain, to move (emotionally), to persuade, to challenge, or even to encourage a behavioral response (such as to destroy or return a contaminated food product or to change certain health practices). The who, what, when, where, why, and how questions are always going to be important, and because digital media do not have the same kind of space and time constraints imposed on traditional media, stories can be told with greater depth and wider breadth. If a specific news event occurs concerning conflict in the Middle East, for example, a fuller "package" of information can be arranged around that specific story so that the reader who needs or wants more content can get it. Obvious things this package might include are hypertext links to other relevant articles, a clickable map of the region with demographic data about the areas in conflict, a historical timeline pinpointing and describing past conflicts in this area, a photo gallery, a slide show with a series of photos and text accompaniments, audio and video clips, and other features. Quick and easy access to archived news stories is clearly a value-added component of digital media. These kinds of multimedia packages, with heavy layering of information options,

address a commonly heard criticism of traditional news as lacking
context and completeness. But even with all these bells and whistles,
the story still needs to inform and explain and to keep the audience
interested through the elements of good storytelling.

Critics of digital media understandably worry that "real news"—
the substance—may get lost in the technological ornamentation or
in the morass of "too much information." This does not have to be
the case and, in fact, does not seem to be the case on most serious
news sites on the Web. Architects of the digital media environment
are learning to organize and present news and information in attrac-
tive, useful, and relevant ways. It is up to journalists, editors, publishers, and their media organizations to maintain professional standards of journalism in this environment, and journalism students
should continue to be taught the fundamentals of good and ethical
writing regardless of what medium they choose to work in. At the
same time, up-and-coming journalists can no longer afford to pigeonhole themselves into a narrow definition of journalism. The
industry is increasingly looking for versatile journalists who can understand, if not work, across a number of different media or in a
convergent media system. The journalist who has an open mind for
learning new things and can adapt to a rapidly changing media environment will be a valuable asset to any media organization in the
years to come.

Journalism students should also ride the digital media wave with a
critical eye and a dose of healthy skepticism. The following questions,
taken and modified from a booklet called "10 Things You Should Know
About New Media" (Kawamoto 1997), can help get students thinking
about the role of digital journalism and digital media in society:

1. What are the advantages of customized news? What are the dis-
   advantages? Could the increase of customized news actually result in a less informed citizenry? How so? How might it result in
   a more informed citizenry?
2. What kind of things can be done to promote more equitable access to new communication and information technologies such
   as the Internet? Is the so-called digital divide still a problem? If
   so, how might it be addressed and resolved?
3. How much of a role should government play in regulating the

Internet? What aspects of the Internet, if any, should government monitor and have some jurisdiction over?

4. What are some effective ways of keeping up with changes in the media environment? How does one learn more about the opportunities and challenges in the field of digital journalism?
5. Should the traditional media (such as print newspapers and magazines or broadcast television) feel threatened by digital media?
6. Is there such a thing as having too much information? Explain your answer.
7. The news media can play a role in building communities and advocating civic engagement. How can digital journalism play a helpful role in this regard?
8. What are some funding models for supporting the digital news media? Can digital media survive financially without a viable funding model? Or is there value to maintaining digital media even though it is not currently profitable?
9. A lot of people have talked about how anyone can be a publisher on the Internet. Why is this a good thing? Are there drawbacks or caveats to having quick and easy access to so-called global publishing?
10. Will the profession of journalism change in the next five to ten years? If so, how? If not, why?

These questions are meant to initiate some discussion about digital journalism and digital media in society. They have no hard-and-fast answers. The goal of these questions, as of the book as a whole, is to encourage dialogue among students, teachers, journalists, and others about the role of journalism in the Digital Age.

Notes

1. Riley et al. write that in the "early 1990s, only a half dozen major newspapers and about a dozen smaller papers had a significant newspaper product or an interactive/online paper on the Web or an Internet provider like America On Line." Patricia Riley et al. "Community or Colony: The Case of Online Newspapers and the Web," *Journal of Computer Medicated Communication* 4(1) September 1998, at http://jcmc.huji.ac.il/vol4/issue1/keough.html.

2. See David E. Carlson, "David Carlson's Online Timeline," at http://iml.jou.ufl.edu/carlson/frames.htm for a comprehensive historical

overview of online news development.

3. Carl Sullivan, "Newspapers Own Half of Top 20 Web Sites," *Editor and Publisher*, July 26, 2002, at www.mediainfo.com/editorandpublisher/headlines/article_display.jsp?vnu_content_id=1559407.

4. See www.ojr.org/ojr/aboutojr/1015887483.php for a full statement of purpose.

5. See www.onlinejournalismawards.org/pr-2001winners1.html for a list of winners with links to their Web sites.

6. A personal digital assistant, or PDA, combines computing, electronic organizing and management, telephone, fax, and Internet features. Users input data with a stylus or mini keyboard or via voice-recognition software. Most people are familiar with the Palm Pilot brand.

7. More detailed information about these services can be found at www.seattlepi.com.

8. See www.indymedia.org/about.php3 for a broader description of this organization.

9. Steve Kettmann, "Net Event Shuts Out Web Press," *Wired News*, February 24, 2000, at www.wired.com/news/politics/0,1283,34551,00.html.

10. Mary K. Feeney, "Olympic Flame: IOC Shuns Internet Journalists," *The Harford Courant*, August 23, 2000, A1.

11. Keith L. Alexander, "Respect Eludes Net Media: Online Reporters Fight for Recognition in Growing Pool," *USA Today*, June 6, 2000, at www.usatoday.com/life/cyber/tech/cth614.htm.

12. Adapted from the Streaming Media West 2002 Press Registration Policy home page.

13. Jane Black, "No, the Net Isn't Annihilating Newspapers," *Business Week*, February 1, 2001, at www.businessweek.com/bwdaily/dnflash/feb2001/nf2001021_527.htm.

Bibliography

DeFleur, Margaret. *Computer-Assisted Investigative Reporting: Development and Methodology*. Mahwah, N.J.: Lawrence Erlbaum Associates, 1997.

Garrison, Bruce. *Computer-Assisted Reporting*, 2nd ed. Mahwah, N.J.: Lawrence Erlbaum Associates, 1998.

Houston, Brant. *Computer-Assisted Reporting: A Practical Guide*, 2nd ed. New York: Bedford–St. Martin's Press, 1999.

Kawamoto, Kevin. "10 Things You Should Know about New Media." San Francisco: The Freedom Forum, 1997.

Kawamoto, Kevin. *Media and Society in the Digital Age.* Boston: Allyn and Bacon, 2003.

Meyer, Philip. *Precision Journalism: A Reporter's Introduction to Social Science Methods.* Boulder, Colo.: Rowman & Littlefield, 2002.

Miller, Lisa. *Power Journalism: Computer-Assisted Reporting.* Belmont, Calif.: Wadsworth, 1997.

Paul, Nora. *When Nerds and Words Collide: Reflections on the Development of Computer-Assisted Research.* St. Petersburg, Fla.: Poynter Institute, 1999.

Riley, Patricia, et al. "Community or Colony: The Case of Online Newspapers and the Web." *Journal of Computer Medicated Communication* 4(1), September 1998, at http://jcmc.huji.ac.il/vol4/issue1/keough.html.

Stephens, Mitchell. *A History of News.* New York: Viking, 1988.

Other Books on Digital Journalism

Callahan, Christopher. *A Journalist's Guide to the Internet: The Net As a Reporting Tool.* Boston: Allyn and Bacon, 2002.

De Wolk, Roland. *Introduction to Online Journalism.* Boston: Allyn and Bacon, 2001.

Harper, Christopher. *And That's the Way It Will Be.* New York: New York University Press, 1998.

Hilliard, Robert L. *Writing for Television, Radio and New Media: With Info Trac.* Belmont, Calif.: Wadsworth, 2000.

Pavlik, John V. *Journalism and New Media.* New York: Columbia University Press, 2001.

Seib, Philip M. *Going Live: Getting the News Right in a Real-Time, Online World.* Boulder, Colo.: Rowman & Littlefield, 2001.

2

The History of Online Journalism

David Carlson

IN 1970, GASOLINE COST TWENTY-FIVE CENTS a gallon, American boys were fighting in Vietnam, the top show on television was "Bonanza," and the introduction of the IBM personal computer was still more than a decade away.

But the roots of online journalism lie in 1970, when virtually no one outside of Isaac Asimov and a few science fiction writers had even considered the possibility that regular people would someday have computers.

Computers themselves had been around since 1945, when ENIAC, the thirty-ton, room-sized Electronic Numerical Integrator and Computer, was built at the University of Pennsylvania, but such computers were huge, incredibly expensive behemoths, and most people believed demand never would be great. They were seen as giant, all-knowing devices that could handle such incredible amounts of information and calculations that IBM itself once forecast that the total, worldwide demand for computers would be five! Only the largest governments and corporations were expected to ever need or want computers.

It was within this odd environment that online journalism was born, but the inventors of the earliest systems expected people to view the information on television sets, not computers. They used computers to create and store the information, but that information was displayed on television sets through the use of special decoder boxes that either sat on top of the set or were built inside. It wasn't until about 1990—some twenty years later—that news and information began to be delivered primarily to computers.

In the three decades since 1970, only two kinds of online journalism systems have emerged: teletext and videotex. But videotex is a broad category that has changed dramatically over the years and includes four distinct types of online systems. We will call them classic videotex, computer bulletin boards, consumer online services, and the World Wide Web.

Teletext

The first type of digital journalism was invented in Great Britain in 1970, the same year a new Ford Pinto cost $1,995 and movie audiences were watching "M*A*S*H" and "Woodstock" in theaters. It was called *teletext,* and it involved displaying words and numbers on television screens in place of regular programming. The invention was patented by the British Broadcasting Corporation (BBC) in 1971, and while teletext has never been a big hit in the United States, it was and still is very popular in Europe and much of the rest of the world.

Definition: Teletext is a noninteractive system for transmission of text and graphics for display on a television set. The set must be equipped with a decoder box or built-in chip in order to capture and display the teletext information.

The consumer uses a remote control to select various pages of information from menus that are created by the teletext operator. Punching in the number 300, for example, might bring up the sports headlines, and entering 305 might bring up a particular sports story. This makes the system appear to be interactive, but it really is not. All of the so-called pages in a teletext system compose a "set." The set is transmitted continuously, one page after another, over and over again. Imagine a big loop, or circle, of pages being sent one after another. When the user enters a page number, the requested page is captured by the teletext decoder the next time it is transmitted. Then it is displayed on the TV screen. No signal or command is sent from the consumer to the teletext system operator.

Teletext can be sent to the home in three ways: television broadcasts (including satellite transmission), cable television systems, or radio broadcasts, but the last is rare. When sent as a television signal, it can occupy a full channel or be encoded in a small part of the channel along with a regular television program. This tiny part of the channel is called the vertical blanking interval (VBI). (If you've ever mistuned a TV set so that the picture rolls and noticed the wide, black line that appears between frames, you have seen the VBI.) Before the development of teletext, the VBI was an unused portion of the television broadcast signal.

This new invention drew considerable attention from the press and public when the BBC announced it in 1971. A working model was first shown in London in 1973. First called Teledata, the BBC system was renamed Ceefax, a play on the words "see facts," before its commercial launch. A few months later, Britain's Independent Broadcasting Authority announced a competing system called Oracle.

FIGURE 2.1
Ceefax, the British teletext system, circa 1984.

The descendants of both still are operating in Great Britain today, and many other countries around the world bought the British technology or developed their own teletext systems.

The advantages of teletext are many for both consumers and operators. The technology created a new use for the television, a device that was and is practically ubiquitous throughout the world. From the consumer standpoint, the systems are free to use and provide useful news, information, and advertising. No telephone or computer is required, and many television sets sold outside the United States have teletext decoder chips built right in. On the business side, teletext enables broadcasters to generate advertising revenue from a part of the broadcast signal that had gone unused before the new technology became available.

But teletext is not without its weaknesses. One drawback is that news and information must be very brief to fit on teletext pages. Most stories are two or three screens long and total fewer than one hundred words. Another is that graphics are poor, preventing photo-quality display. A third is the amount of time a consumer must wait for a page to display after entering its number in the remote control. This can vary from a few seconds to nearly a minute, which most people find to be an unacceptably long time. The delay depends on several factors, including the number of pages, or frames, in the teletext set and at what point in the transmission of the set the consumer enters the page number. If someone requests page 399 at the moment page 1 is being transmitted, the delay will be much longer than if that request is entered while page 350 is being sent. Most teletext sets are limited to about five hundred pages to keep this delay to thirty seconds or less.

More than twenty-five American companies, including television networks and stations, newspapers, magazines, cable operators, and others, launched teletext systems between 1975 and 1984, but none were commercially successful enough to survive long-term. The last major U.S. system, carried in the VBI of Atlanta's cable superstation WTBS, closed in early 1994.

Videotex

Very soon after the invention of teletext, another type of digital journalism was invented at British Telecom, then a research laboratory operated by the British Post Office. Originally called *viewdata* but later dubbed *videotex*, it was the big breakthrough, the forerunner of all of today's interactive, online systems.

The big breakthrough was that, unlike teletext systems, videotex truly was interactive, allowing consumers actually to communicate with the videotex system to send or receive data.

Definition: Videotex systems are interactive, computer-based systems that electronically deliver text, numbers, and graphics for display on a television set, video monitor, or personal computer. The data travels over telephone lines, two-way cable, computer networks, wireless data networks, or any combination of the four.

Every interactive online system that has existed, including the Internet and the World Wide Web, falls within the following definition of videotex:

- It is interactive, meaning two-way communication is supported.
- Data is stored in computers, often PCs, but sometimes minicomputers or mainframes.
- Users enter commands on a keyboard, dedicated terminal, or computer. Those commands are sent to the host computer, and the requested data is returned to the user.
- Navigation can be accomplished either through menus or command-line interfaces.
- Menu-driven systems allow users to browse, much as they would a newspaper.
- Command-line interfaces allow very fast searches for specific data.
- Graphics are much better than those presented in teletext, eventually even allowing photo display.
- Messaging and bulletin boards—among the first truly interactive, participatory services—are supported.

The latest and most popular versions of videotex systems are AOL and the World Wide Web.

Videotex was not originally delivered to computers. Instead, a television set hooked to a set-top box was used to receive information from a remote database via a telephone line or cable TV. The early services offered thousands of pages ranging from consumer information to financial data, but with limited graphics. Such services were offered in many countries around the world, most notably Great Britain, France, and the United States.

Great Britain

The world's first videotex system was Prestel, created by the British Post Office and eventually offered throughout the United Kingdom. Sam Fedida, computer applications manager at the British Post Office Research Laboratories, was the acknowledged inventor. He was a research engineer who had been at the fore of a host of cutting-edge projects at the labs.

The first demonstration of Prestel took place in London in late 1974 on a Hewlett-Packard minicomputer. After testing in various U.K. markets, British Telecom, the government-owned telephone company, launched it commercially in early 1979, and the world took notice.

Dozens of companies and countries leapt to adapt the new technology and cash in on what was forecasted to be a huge market with home consumers and commercial businesses. A whole range of applications was imagined, and news and information and online banking were among those thought to have the most commercial promise. Newspapers throughout the United Kingdom lined up to deliver their data via Prestel. It's unclear exactly who was the very first to offer newspaper stories online, but the *Financial Times* of London, the *Post and Echo* of Liverpool, and *Eastern Counties Newspapers* were among the first. Their services were launched in 1978 while Prestel was being tested in London.

The information available via Prestel was eventually provided by over one thousand independent sources known as information providers (IPs). It ranged from stock exchanges to the British Consumer's Association and the government Meteorological Office. There were train and ferry timetables, racing tips, best-buy recommendations, consumer advice, dining guides, health-care information, airline schedules, and much more, including news and sports from several sources. Electronic shopping for thousands of services and products also was available.

By January 1983, there were thirty thousand Prestel terminals in use, and Prestel estimated that two hundred thousand people were using the service. Despite initial optimism, however, customers among home consumers proved elusive for Prestel and for most of its imitators and adapters around the world.

The services were too expensive to attract consumers in any significant volume. The cost of the adapted televisions was nearly three times that for a normal set. In addition, there were charges for access to the system, as well as telephone bills. (It is important to understand that even local telephone calls are charged by the minute in most parts of the world.)

When home consumers balked at the cost of using Prestel, British Telecom focused on the business sector and tried to woo consumers with software to access the service via computer. But by the late 1980s, the service still was struggling to make any commercial impact. Even so, success was believed to be just around the corner, and Prestel struggled on for years. It is a sad irony that Clive Fedida, son of the inventor and a British Telecom executive himself, was instrumental in closing Prestel in the spring of 1994, fifteen years after its commercial launch.

Ahead of its time and without the benefit of cheap personal computing, the technology became an expensive failure for the British and many others.

One exception was France.

France

The French Teletel videotex network (its underlying technical system is called Antiope) is arguably the greatest videotex success story of the 1980s. Created in 1980 by France Telecom, the French government's telephone company, it still operates in the early twenty-first century.

From the start, the service was innovative. It replaced telephone books containing every telephone listing in France, and it featured train, plane, bus, and subway schedules, as well as volumes of news and information.

To use it, viewers needed a dedicated Minitel terminal, which initially was given free to telephone subscribers. The early terminals featured a built-in modem (1,200 bps receive and 75-bps send), a fold-up keyboard, and either a nine-inch black-and-white screen or a twelve-inch color screen. The system was later adapted to use computers for access as well as the terminals.

The system essentially was financed by its electronic telephone book, and revenues from Teletel were used to rebuild France's aging phone system. France Telecom used the money it would have spent to publish and deliver conventional phone books to finance much of the construction of the videotex system. Then, revenues generated by telephone usage were used to help finance reconstruction of the phone system itself.

Now, more than twenty million users can access more than twenty-five thousand services on Teletel/Minitel. The service is even available in the United States, but most of the information on it is in the French language.

By 1994, there were 6.5 million Minitel terminals and six hundred thousand microcomputers in France equipped with the appropriate software. The system was generating over two billion calls a year from users.

According to France Telecom, "Minitel in France provides a wealth and diversity of services unrivaled elsewhere. It meets the needs of all users, residential and business alike."

United States

France's incredible videotex success story attracted a great deal of attention, especially among American newspaper companies. Many of them scrambled to create and introduce videotex services. But it is important to understand that the efforts in France and Great Britain were government owned and operated. In both cases, the government telecommunication companies, which ran the telephone systems, developed the videotex systems, and most of the online information was provided by independent, private companies. It was different in the United States. The services largely were developed by newspapers and broadcasting companies in partnership with technology companies such as AT&T and Honeywell Corporation.

More than twenty-five newspaper videotex services were introduced in the United States between 1981 and 1986. Perhaps most notable among them were Viewtron, a Knight Ridder initiative, and Gateway from Times Mirror's Los Angeles Times Company.

Viewtron

Perhaps the most ambitious videotex trial in the United States, Viewtron was a joint venture of Knight Ridder, AT&T, and other minor partners, including Scripps Howard and The McClatchy Company. The partners formed Viewdata Corporation of America, which began market trials in Coral Gables, Florida, an upscale suburb of Miami, in 1980. The commercial service was launched in South Florida (Dade, Broward, Monroe, and Palm Beach counties) on October 30, 1983.

The system emphasized consumer products, such as weather information, banking, and shopping. News from Knight Ridder's *Miami Herald* and the Associated Press was the heart of the service. Eight local schools provided neighborhood news. Other items included local sports, police reports, news from local colleges and universities, neighborhood calendars, and biographies of prominent local citizens.

Viewtron was displayed on television sets through the use of Sceptre terminals, set-top boxes that were manufactured by AT&T and included a built-in modem and an infrared keyboard. Data was sent to the home via cable television, and users sent commands back to the system via modem and telephone line.

Pricing was a problem. To use the service, viewers had to buy the Sceptre terminal. At launch, they cost $900 and were reduced to $600 when demand was soft. Further, a subscription in Miami cost $12 a month, plus long-distance phone charges if used from outside the core dialing area. This was at a time when most consumers could have a daily newspaper delivered to their doorstep for $5 a month. There also were additional charges for some services, such as e-mail from Hallmark Cards at $2 per card.

The company had predicted five thousand subscribers by the end of the first year of operation. Only half that number was achieved with the company recording some $16 million in losses.

After May 1984, the partners gave up trying to sell the Sceptre terminals and changed the pricing system to $39.95 a month including terminal rental. They also began to develop and market software that would allow personal computer owners to use Viewtron.

Despite its meager success in the Miami area, Viewtron expanded to include all of Florida in 1984 and other U.S. cities by 1985. At its height, Viewtron was operated in at least fifteen cities by various newspaper companies. Knight Ridder had operations in Miami, Philadelphia, San Jose, Detroit, Charlotte, and St. Paul, and other versions of Viewtron were operated in Baltimore, Boston, Fort Worth, Kansas City, Seattle, New Orleans, Portland, Cleveland, and Newark by other newspaper companies, including Newhouse and Capital Cities.

After an investment reportedly in excess of $55 million by Knight Ridder and $10 million each from Scripps Howard and McClatchy, Viewtron closed in 1986.

Said to be one of the three most comprehensive videotex trials conducted in the United States, the Southern California Field Trial of Gateway ran from March 15 to December 31, 1982. The service then operated commercially from 1983 to 1986.

The trial involved some 200 households in Ranchos Palos Verdes and 150 in Mission Viejo and was conducted by Videotex America, a joint venture of Times Mirror Videotex Services, a subsidiary of the Los Angeles Times Company, and Infomart of Canada. The services used the same Sceptre terminal required by Viewtron and manufactured by AT&T.

The four most useful services, as ranked by the testers, were news, banking, shopping, and games. During the trial, Gateway offered twenty-one main index categories, and news/weather ranked consistently among the top two or three. Games, e-mail, and bulletin boards were the other most popular categories. The service began the trial with some twenty thousand pages of information and had grown to seventy thousand pages by the end.

While many details of the user studies were kept proprietary, participants were reported to have said they wanted a comprehensive home-information service and not just a newspaper or entertainment system. They also said they expected videotex news to be updated constantly, easily retrieved, available on demand, and, whenever possible, interactive. They said they expected videotex to supplement their daily newspaper, however, not replace it.

Gateway went commercial in the fall of 1984 and was predicted to break even by 1988 or 1989, but with pricing similar to that of Viewtron, consumer demand again was soft. It closed in late 1986 with the partners having invested an estimated $20 million.

At its height, Gateway technology was used by newspaper companies in seven U.S. cities: Los Angeles (Times Mirror), Minneapolis

(Cowles Media), Jacksonville, Florida (*The Jacksonville Journal*), Sacramento (The McClatchy Company), Phoenix (Pulliam Newspapers), San Francisco (*San Francisco Chronicle*), and Washington (*Washington Post*).

What Went Wrong?

A lot of arguments could be made about why all the early newspaper-operated videotex services failed to gain enough market share to survive. Let's look at the things they had in common that likely contributed to their demise:

- *Dedicated terminals were required for access.* These dedicated terminals proved difficult to market because they cost as much as $900 to buy or $30 per month to rent. In defense of the system designers, the terminals had been expected to cost less than $300, but they turned out to cost more than that to manufacture. The terminals also were single-use machines that might not even work with a competing system in another city. A Sceptre terminal bought for Viewtron in Miami would not work, for example, with Keycom in Chicago, which required a different terminal manufactured by Honeywell. Further, these terminals came to market at about the same time the public became aware of the first personal computers from Apple and IBM. (The Apple I was introduced in April 1976, but the first real success was the Apple II in 1978, followed by the Macintosh in January 1984. The IBM PC debuted in August 1981 and sold fifty thousand units in the first eight months. Less than two years later, the first Macintosh sold fifty thousand units in its first seventy-five days.)
- *They tied up the family TV and the telephone.* This may have been the largest single factor in the failure of the early videotex systems, and it was a problem largely overlooked in test-marketing trials. Few American families in the early 1980s had more than one television set, or more than one phone line, and using an online service did not prove to be a family activity, such as gathering around the TV after dinner to watch the evening sitcoms. It is a solitary activity. Each family member is likely to have different interests and

different favorite online applications. A parent may want to do banking and check investments, while a child may prefer to play games or trade e-mail with friends. But with the family TV tied up as the videotex display device—and the telephone in use to send information to the videotex system—the rest of the family had little to do for entertainment. This proved unpopular, but test-marketers for Viewtron, for example, did not discover the problem because they had given the families participating in the trials an extra TV and telephone line.

- *Costs were high for the times, and pricing schemes were complicated.* The average charge for use of a videotex system was $19.95 a month for basic service (incidentally, about the same amount Internet service providers still charge for unlimited usage early in the twenty-first century). But usage was not unlimited. There were often per-minute charges over and above the basic rate. There also were additional costs for access to specialized information such as stock quotations or airline reservation systems. The result was an average bill of some $30 a month at a time when newspapers were delivered every day for about $60 a year. User studies showed that consumers often were confused about exactly how much they were being charged for certain types of information, which made them uncomfortable. Further, the systems were extremely expensive to build and operate. Mainframe computers costing millions of dollars were required to house the systems, and software had to be built from scratch.

- *They had poor messaging capabilities.* Amazingly, none of the designers of these early videotex systems anticipated the demand that the introduction of e-mail would create. The infrared remote keyboards of the Sceptre and Honeywell terminals were small so they would be easy to use from across the room, but that meant they had tiny keys that were difficult to use for conventional touch-typing. Even so, consumers quickly discovered how convenient e-mail could be and sent it by the boatload, clogging early videotex systems, which had not been designed to handle the massive volume of messages. This definitely was a harbinger of the future. Electronic messaging has been the single most popular feature of every online system so far created, including the Internet.

Bulletin Boards and Consumer Online Services

The result of all this activity was that American newspaper companies collectively had lost at least $100 million on videotex by 1986. Nearly all of them threw in the towel and closed their systems. There were but two notable exceptions: the *Fort Worth Star-Telegram* with a service called Startext and a consortium of IBM, Sears Roebuck and Company, and the CBS television network, which in 1985 began developing a service it called Trintex. It was renamed Prodigy before it finally launched in 1988.

The difference between the classic videotex services and these two was an important one: Startext and Trintex were intended from the beginning for delivery to personal computers—not television sets.

They were part of a development parallel with classic videotex that has come to be called consumer online services. These were designed from the beginning to be delivered to home computers, and, while a great deal of literature refers to them as commercial online services, this really is a misnomer. Commercial online services actually are a completely different class of services devoted to commercial activity such as stock trading and databases of publications, legal decisions, and medical information. When you use an automatic teller machine to withdraw money from your checking account, you are using a commercial online service.

Both consumer and commercial online services actually date back to the 1960s, even preceding the development of teletext. But they really did not become popular until the mid-1980s when personal computers began to appear widely in offices and homes.

While we will not devote a lot of attention here to commercial online services, the first of them appears to have been Medlars, the Medical Literature Analysis and Retrieval System, which was launched in 1962 as a service for hospitals and doctors. Another one, Dialog, came along in 1969 as a provider of databases of publications, especially back issues of trade, professional, and academic journals. It survives today.

The first consumer online service was CompuServe Information Service, founded in Columbus, Ohio, in 1969. It was the brainchild of Jeffrey M. Wilkins.

Wilkins's father-in-law, who ran an insurance company, needed to buy a computer. He wanted a particular model from Digital Equipment

Corporation, but the only one available turned out to be much larger than he needed. Then, the two men had an idea: Why not buy it anyway and rent out the extra capacity?

Wilkins quit his job in the burglar alarm business and set about running the new company. Initially, CompuServe sold its excess computer capacity to other corporations, but in 1978 it began providing services to the owners of personal computers. The goal was to squeeze profits out of underutilized assets by putting time-sharing computers to greater use at night, when they were frequently idle.

While CompuServe was not a smashing success in its early years, it and a competitor, The Source, had garnered some fifteen thousand user accounts by 1982. This prompted other entrants in the consumer online service sector.

One of these, Startext, was operated by a newspaper and began in 1982 as a partnership between the *Fort Worth Star-Telegram* and Tandy Corporation, known primarily as the operator of RadioShack stores. Tandy was a fairly early entrant into the personal computer marketplace with several models of home computers that were sold at RadioShack. In 1983, it introduced the first portable machine to really catch the fancy of journalists, the RadioShack Model 100, which ran on four AA batteries and had a forty-character by eight-line, black-and-white LCD screen. The machine, with built-in modem and software for word processing, quickly became a favorite of reporters.

The *Star-Telegram* and Tandy agreed that Dallas–Fort Worth would be a fine test market for an online service utilizing Tandy computers, and they worked together to create Startext. The system was never profitable, and Tandy eventually dropped out, but the *Star-Telegram* kept it operating throughout the 1980s and even into the mid-1990s, when it finally moved its news services to the World Wide Web.

Startext was housed on a mainframe computer. That computer was connected to a bank of telephone modems, devices allowing computers to send and receive information over telephone lines. Startext consumers would install a modem and software package on their personal computers and use them to dial the phone and connect with the service at a cost of about $9.95 a month for unlimited access.

But Startext was not particularly interesting for its technology, which was a very basic, command-line interface. It was interesting for its content.

The *Star-Telegram* did not just post its own news and information online. It encouraged Startext users to contribute content to the online service. There were user-contributed movie reviews, commentary, columns, short stories, and more, all of which helped to build an active online community and an interesting and useful set of information.

In the late 1980s and early 1990s, the newspaper tried to market the Startext operating system to other newspaper companies, just as Viewtron and Gateway had done years before. The *Post-Dispatch* in St. Louis tried the system in 1992, but that service closed by 1994.

Meanwhile, the other consumer online services were experiencing dramatic growth by the early 1990s. CompuServe was passing the one-million-subscriber mark. Prodigy, available in test markets since 1988, launched nationally in 1990 and had five hundred thousand subscribers within a month, and AOL, originally called Apple Link, a service for Apple computer owners, was developing software for PC users.

Newspapers, however, still were striking out on their own. The next wave of online journalism came through the use of computer bulletin board systems, or BBSs.

Invented in 1978, BBSs are software packages that enable a personal computer to house a complete interactive online system. The computer, hooked to one or more modems, answers calls from consumers' computers and offers them access to news, information, e-mail, discussion areas, and even teleconferencing in which a group of people can type messages and interact with one another in real time. BBSs originally could be accessed only via computer modem, but later, some were accessible through computer networks such as the Internet.

In the late 1980s some people in the newspaper industry began to see BBSs as a potential way to offer an inexpensive videotex system. IBM PCs were becoming reasonably powerful and cost thousands of dollars rather than the millions mainframes used by classic videotex systems cost. Coupled with an over-the-counter BBS package, a reasonably robust online newspaper could be created.

At least thirty U.S. newspapers launched computer BBSs between 1990 and 1994. Most of them were smaller regional dailies, rather than the large newspaper groups and metropolitan dailies that had ventured into classic videotex.

The pioneer of large BBSs run by newspapers was The Electronic Trib, a service launched in 1990 by the *Albuquerque Tribune* in New

FIGURE 2.2
A BBS service called Sun.ONE, a partnership of the University of Florida Interactive Media Lab and the New York Times Regional Newspaper Group, May 1995.

Mexico. The system, created for a capital outlay of just $5,000, ran on an IBM-compatible computer with a 286-12 MHz processor. The E-Trib, as it came to be called, was unlike virtually any other newspaper online system of any period, and it achieved a number of firsts.

For one, it was not conceived with the idea of making money. Access was free for thirty minutes a day with a password published in each day's newspaper. The *Tribune* only began charging for access after readers requested more time and offered to pay for it. For another, most of the E-Trib's content was not the same as that published in the newspaper. Contending it wanted to supplement the daily newspaper, not replace it, the E-Trib included some news from the *Tribune*, but mostly it carried stories that had not been in the paper, such as the complete daily feed from the Scripps Howard News Service. Other firsts included offering online databases of public records, holding chat sessions with newsmakers, public officials, and editors, and providing access to Internet e-mail. E-Trib editors even figured out a way to let online users order pizza for home delivery by 1992.

As adventurous smaller newspapers were building BBSs, some larger newspapers were forging partnerships with the consumer online services. CompuServe, Prodigy, and AOL all carried various newspapers' content between 1993 and 1996.

CompuServe was the old salt, having actually hosted newspaper content as early as July 1980 when the *Columbus Dispatch* began offering its material experimentally. The *Dispatch* was joined by ten other papers, including the *New York Times, Washington Post,* and *Los Angeles Times,* during a short-lived experiment in 1981 and 1982.

A new wave began in November 1991 when the Tribune Company of Chicago announced it would offer a service on AOL that would include content from all of its Chicago media properties, including radio, television, and its flagship newspaper, the *Chicago Tribune.* AOL was by now reinventing itself as a service for PCs running the Microsoft Windows operating system.

Chicago Online's launch in May 1992 was followed by a flurry of other services on each of the big three consumer online services. Gannett's *Florida Today* put a space-oriented service on CompuServe, and Cox Newspapers announced it would partner with Prodigy to put the

FIGURE 2.3
Prodigy service, circa 1994. The image on the right is a "pop-up photo," a feature Prodigy was beta testing at the time.

Atlanta Journal and Constitution and the *Palm Beach Post* on the Prodigy service. The *San Jose Mercury News,* a Knight Ridder paper, launched a service on AOL in May 1993. *Time Magazine,* the *New York Times,* and others followed with AOL services, and still others, including *Newsweek* and *Consumer Reports,* went with Prodigy.

By the fall of 1993, the big three consumer online services had a combined 3.9 million members. CompuServe and Prodigy were the largest, but AOL had the best interface. As they continued to grow, and newspapers and magazines around the country forged partnerships with them, a new, little-known piece of software was released by the University of Illinois that August. It was called Mosaic, and it would change everything.

The World Wide Web

Mosaic was something called a browser, and it put a graphical interface on top of an already existing but virtually unknown Internet service called the World Wide Web.

The Internet had existed for a long time by 1993. It was developed in the 1960s by the Advanced Research Projects Agency, an arm of the U.S. Department of Defense, but few people outside the military, major universities, and defense contractors even knew of its existence. Access was limited to noncommercial traffic, and the interface was confusing and completely nongraphical.

Mosaic was the catalyst that made the Internet really take off—and eventually caused newspapers, and virtually everyone else, to focus their online efforts on the World Wide Web.

To understand the phenomenon that is the Web, it is important first to understand that the Web and the Internet are not interchangeable terms. The Internet is a vast, global network of computer networks that are able to work together seamlessly through the use of a common language, or protocol. Just as telephone systems in various nations can communicate because they use similar technology, the Internet Protocol enables computers of different manufacture all around the world to communicate and work together.

That's all the Internet is—a network of networks. But many different services have been developed over the years to make the Internet more useful. Perhaps the best known and most popular of them is e-mail. Others include File Transfer Protocol (FTP), which allows computer files to be moved around; TELNET, which enables remote users to log into computers elsewhere on the network; and, the best known of all now, the World Wide Web. So, the Web, TELNET, e-mail, and FTP all are services that exist on the Internet.

Many people think the Web and the Internet are interchangeable terms because all the major Internet services now can be used through a single software package, the browser. You don't receive your e-mail over the Web. It just seems as if you do because your browser includes software that can send and receive your mail. You don't download a new piece of software over the Web either, but it might seem as if you do because your browser includes FTP software.

Mosaic was the first graphical Web browser for Microsoft Windows, but it was not the first browser. Several came before it, including Lynx, a program that offered text-only navigation of the Web for PCs, and Viola, a browser for computers running the UNIX operating system. None of them, however, attracted the attention Mosaic garnered, mostly because Windows was by then the operating system of choice for most home and office computing.

The World Wide Web itself was created in 1990 when Tim Berners-Lee, an Englishman, and colleagues at the European Center for Particle Physics (CERN) developed the computer language that enabled users to navigate by simply clicking on underlined words called links. The language, called Hypertext Markup Language (HTML), is still the basis of Web pages today.

Berners-Lee first called the resulting linkage of various computers the World Wide Web in a paper he wrote in October 1990. He later recalled that other names he considered were Mine of Information and Information Mine.

Growth at first was slow. In November 1992, CERN estimated that twenty-six "reasonably reliable" servers existed on the Web. But growth skyrocketed after the release of the Mosaic browser, which made the Web accessible to millions of people. It was estimated that Mosaic spawned an annual growth rate of 341,634 percent in late 1993! Newspapers and journalists were quick to notice.

The first journalism site on the Web launched in November 1993 at the University of Florida College of Journalism and Communications, and, on January 19, 1994, the *Palo Alto Weekly* in California became the first newspaper to publish regularly on the Web. Its full content was posted twice weekly, and access was free. That same month, *E&P* reported that twenty online newspaper services existed worldwide, most of them BBSs.

Other early Web newspapers in the United States were the *Gazette-Telegraph* in Colorado Springs, Colorado; The Electronic Signpost at the *Star-Tribune* in Casper, Wyoming; and Pilot Online at the *Virginian Pilot* in Norfolk, Virginia. In July 1994, the *News and Observer* in Raleigh, North Carolina, added NandoTimes and the SportsServer while still operating a BBS system. In September, Time Warner became the first big media company on the Web with Pathfinder, a site that offered content from all its major magazines, including *Time, People,* and *Money.*

Some newspapers and magazine publishers quickly recognized that the Web offered an interface closer to the newspaper model than any form of videotex had before. For the first time, it was possible to put headlines, photos, captions, and text together on a single page in a way that closely represented the classic look of a newspaper. Others balked at the idea because, at first, there was no potential for revenue because there was no way to charge Web users for access and the audience was too small to interest advertisers. But some publishers were quick to recognize the potential, and soon the rush was on.

Three events in the latter half of 1994 were major catalysts. First, the *San Jose Mercury News,* recognized as a major success with its AOL service called Mercury Center, announced it would add a Web site to its online ventures. Then, Prodigy and AOL announced they would include a Web browser in their software packages. Finally, on November 1, a major newspaper strike closed down both of the major San Francisco dailies. Strikers and management created rival newspapers on the Web, the *Free Press* and the *Gate.* These attracted considerable attention from the public and newspaper executives around the world.

Even as these events were unfolding, though, large newspaper and magazine companies were still partnering with the consumer online services. The *New York Times* launched a service called @Times on AOL while the *Los Angeles Times* and *Newsday* created sites on Prodigy. The *Washington Post* and *Minneapolis Star Tribune* were working with a new online service called Interchange, and even Microsoft was working to create a new consumer online service to be called The Microsoft Network.

The reasons were simple. The American public had begun to discover the online world, and the consumer online services were experiencing phenomenal growth, 64 percent in 1995 alone. Together, they reached 8.5 million subscribers by the end of 1995. For newspapers, this meant a large audience and immediate revenue because the online services shared the fees they charged their members with each of their information providers.

By May 1995, some 150 U.S. dailies had some type of online service, *Quill* reported. More and more were moving to the Web, but many still were hesitant because less than 1 percent of the American population yet had access to the World Wide Web. There was no good means to charge users for access to Web-based newspapers—something all previous online services had done—and there were few advertisers interested in such a small audience. This made revenues scarce.

This began to change, though, as the government prohibition of commercial activity on the Internet was dropped, and first Prodigy, then AOL, and finally CompuServe offered their subscribers access to the World Wide Web. This increased the Web audience almost overnight—and made it reasonably possible for newspapers to generate revenue from the sale of advertising on their Web sites.

In the last six months of 1995, the number of online newspapers more than doubled. *E&P* reported somewhat breathlessly in November that 330 newspapers in the United States either were online or had announced projects. It said 38 operated BBSs, 45 were affiliated with Prodigy, AOL, or CompuServe, and 230 were on the Web or intended to be. These figures probably were slightly inflated as the Newspaper Association of America reported the following April that 775 newspapers were on the Web worldwide, 175 of them in North America.

The next year proved to be the biggest yet. In January, the *New York Times* launched its Web product, the *New York Times* on the Web. Then came the *Chicago Tribune*, the Associated Press, the *Wall Street Journal*, the *Washington Post*, and many more. In July 1996, the *San Jose Mercury News* became the first to announce it would drop its consumer online service partnership and focus solely on Web-based ventures.

Thus began the exodus of publishers from the consumer online services. By 1997, nearly all of them were focusing on the Web, although several still operated sites on the consumer services. Even without newspapers, though, the consumer services were growing by leaps and bounds. In April 1997, they boasted 16.3 million members: 8 million on AOL, 5.3 million on CompuServe, 2 million on the Microsoft Network, and 1 million on Prodigy. Their growth, however, was not so much related to their proprietary content as it was to their providing access to e-mail and the Web.

By 1999, AOL had grown to fifteen million members and bought CompuServe. Microsoft and Prodigy had stopped operating as proprietary online services, and the number of newspapers on the Web had grown to an estimated fourteen thousand around the world. Even so, fewer than 50 percent of them were estimated to be profitable.

That lack of profitability would catch up to them early in the next century. Throughout the late 1990s, the stocks of Internet companies soared, even though there were few profitable online sites, and all kinds of companies, including media, invested massive amounts of capital in creating Web sites. Speculation and optimism for the future of any and all companies involved in Internet commerce was wild. The *New York Times* and the *Washington Post* each had nearly two hundred people on their Internet staffs. AOL's stock climbed so high that in January 2000 it was able to buy Time Warner in a deal estimated to be worth $162 billion.

When profits failed to materialize, though, Internet stocks began to fall precipitously in the early 2000s. Stock in Yahoo!, an extremely popular Internet portal, reached $216 a share in December 1999, and Yahoo! alone had market capitalization greater than the six largest American newspaper companies combined. By March 2001, though, Yahoo! stock had fallen all the way to $11.37.

Even as online companies announced massive layoffs and many small startup companies failed, the number of people using the Internet was continuing to grow. In May 2001, AOL claimed to have twenty-nine million subscribers. The *New York Times* reported that month that 57 percent of American households had some type of Internet access, and the *Wall Street Journal* said some 461 million people worldwide were connected to the Internet. Later that year, the *Journal* said 68 percent of

American households had computers, and the vast majority of them had Internet connection available.

Conclusion

You may wonder after reading this chapter why so many newspapers and media companies are involved in online journalism and why you might want to be involved.

The reason is not hard to figure out: its potential.

Practically from the first demonstration of Viewdata in 1974, newspapers and others have recognized its possibilities. Online journalism has the potential to take the best things about newspapers, magazines, radio, and television and wrap them up into a single new product with all the strengths of existing media and few, if any, of the weaknesses.

An online journal can have the depth and breadth of newspapers, the immediacy of radio, television's ability to make us "eyewitnesses" to events, and the typography and resolution of slick magazines.

Imagine a wireless, portable-computing device on which your customized daily newspaper is displayed. At the top of your front page is a story about how your personal wealth has changed in the past few minutes or days as a result of the world's financial markets. That's much more interesting than the Dow Jones Industrial Average. Now, click on the picture of the umpire making a call at the plate in a baseball game and see the full-motion video, just like on the TV news. But with this device, you can play it backward or forward in fast motion or slow and decide for yourself if the call was correct. Then, click on the box score and find out you have access to every statistic about every game ever played between these teams and every player who ever played for them.

It sounds pretty compelling, doesn't it? But that's only part of the story. To really understand why media companies are so interested in online journalism, we must also talk about money.

On average, eighty cents of every dollar newspapers spend goes to production and delivery of the daily paper. This reflects the cost of presses, paper, ink, trucks, fuel, and the personnel who print and deliver the newspaper—all costs that are practically eliminated in online journalism. Presses alone cost millions of dollars—enough that most local newspapers could give every subscriber a portable computer for less than the cost of buying a new press.

The other side of this coin is that, again on average, about twenty cents of every dollar newspapers collect comes from circulation revenue—the money people pay to buy the paper. This revenue is lost in the online world, at least in most current economic models.

Add it up, and you see why newspapers and other media are so excited about online journalism: They can create a new and better product, give it away for free, and still save a lot of money, potentially increasing their profitability.

That, in a nutshell, is why newspapers have been involved in videotex since its earliest days. Along the way, a number of lessons should have been learned:

- *To be successful, online services, news-oriented or otherwise, must be personally useful.* From the earliest days of videotex, this lesson was made obvious by the instant popularity of e-mail. Even today, the most popular Web sites are portals—sites that provide an entryway to the Web—auction sites, and search engines, all of which are personally useful.
- *Interactivity is a key element.* The biggest weakness of the traditional mass media is their lack of interactivity and two-way communication. It is very difficult for readers or viewers to interact with one another or with reporters and editors. This limits feedback that lets media executives know how good a job they are doing. Online journalism can overcome this weakness easily since the medium itself is interactive.
- *Content must be free unless it is very specialized.* Several publications have tried to charge for access to their online content. Only those with extremely specialized material not available anywhere else have been successful. The *Wall Street Journal* has been reasonably successful selling subscriptions to its WSJ.com Web site for $39 to $69 per year, but virtually all others who have tried it, including the *New York Times,* which once charged overseas users of its Web site, have gone back to allowing free access.
- *The real money is not in selling technology. It's in selling programming.* The inventors of new technology—from the telephone to the

radio, the television set, the videocassette recorder, the compact disc player, and even the videotex terminal—usually have thought the money would be made from selling the equipment, the hardware. Without fail, the biggest profits have ended up coming from the sale of programming for the new medium, the software. The money generated from the sale of radio sets, for example, hasn't come close to the money generated from the sale of advertising on radio.

This same lesson is beginning to be learned on the World Wide Web. Many newspapers were hesitant to adopt the new medium because they would be giving their content away (just as radio stations always have). Some are discovering that giving their content away is bringing readers in sufficient quantity to attract advertisers who will pay to reach those readers.

3

The Meanings and
Implications of Convergence

Rich Gordon

A T THE DAWN OF THE TWENTY-FIRST CENTURY, it seems nearly impossible to follow developments in technology, business, or journalism without encountering the word *convergence*. The term has been applied to corporate strategies (the merger of AOL and Time Warner), to technological developments (video on demand and interactive television), to marketing efforts (partnerships between newspapers and TV stations to promote each other's work), to job descriptions ("backpack journalists," who return from the scene of a story with words, audio, and video), and to storytelling techniques (the melding of text and multimedia on news Web sites). There is a danger that "convergence" will become a buzzword, thrown around casually in discussions of media, technology, and journalism, with different participants using the same word to mean different things. If we are going to think clearly about journalism's present and future, we're going to need to understand these different meanings and their implications.

Tracing the Meanings of a Word

The term *convergence* comes originally from the world of science and mathematics. According to the Oxford English Dictionary, its earliest use can be traced to William Derham, an English scientist in the seventeenth and eighteenth centuries who is best known for his effort to measure the speed of sound by timing the interval between the flash and roar of a cannon. In 1713, Derham's *Physico-Theology: Or a Demonstration of the Being and Attributes of God, from His Works of Creation* referred to "convergences and divergences of the rays." In the ensuing centuries, the term was applied to wind currents, mathematical series, nonparallel lines, and evolutionary biology. (Charles Darwin used the term in the 1866 edition of *The Origin of Species.*)[1] By the middle of the twentieth century, the term was also being applied to political science (convergence of U.S. and Soviet systems) and economics (convergence of national economies into a global economy).

In the 1960s and 1970s, the development of computers and networks established the context for new meanings. Government agencies and businesses began using computers to store and retrieve information. Then, they began transferring this digital content over telecommunications systems. Finally, in the mid-1970s, came the realization that once information could be stored digitally and communicated across a network, the technologies that consumers use to access information and entertainment could be transformed. Commercial enterprises began to experiment with computer online services and videotext delivered to TV screens. The first online services in the United States, The Source and CompuServe, began offering dial-up access in 1978. At about the same time, the British Post Office was rolling out its Prestel videotext service.[2]

It won't be easy to figure out exactly who first used the word "convergence" in connection with communications technologies. But communications scholar Ithiel de Sola Pool clearly helped popularize it. In his landmark 1983 book, *The Technologies of Freedom*, Pool described what he called "the convergence of modes":

> The explanation for the current convergence between historically separated modes of communication lies in the hability of digital electronics. Conversation, theater, news, and text are all increasingly delivered electronically. . . . [E]lectronic technology is bringing all modes of communications into one grand system.[3]

Even before Pool's book was published, leading thinkers in journalism and academia were coming to realize that technological changes were going to affect the news media. William Paley, chairman of CBS, gave a speech to a broadcasters' convention in 1980, noting that "the convergence of delivery mechanisms for news and information raises anew some critical First Amendment questions."[4] And as early as 1979, Nicholas Negroponte of the Massachusetts Institute of Technology (MIT) was using three overlapping circles in discussions with business executives he hoped would fund his research. His three circles were labeled "Broadcast and Motion Picture Industry," "Computer Industry," and "Print and Publishing Industry."[5] He predicted that the overlap between the three circles would become almost total by 2000. The executives he addressed found it a compelling vision. He won millions of dollars in financial support from them, enabling MIT to open its celebrated Media Lab in 1985.

One of those who came to embrace Negroponte's view of the world was John Sculley, who left a top executive position at Pepsi in 1983 to become CEO of Apple Computer. To illustrate his vision for Apple in the future, he would display two graphic representations of the "information industry" in 1980 and 2000. For 1980, the graphic showed seven boxes, each representing a discrete industry: media/publishing, information vendors, computers, consumer electronics, telecommunications, office equipment, and distribution. For 2000, the graphic was labeled "Convergence" and represented the industries as overlapping ellipses. And there were new labels where the shapes overlapped, such as "Interactive News," "Virtual Reality," "National Data Highway," "Info on Demand," and "2-Way TV."[6]

When the word *convergence* showed up in the business press in the 1980s and early 1990s, it was often in connection with Sculley and Apple. The word also appeared in articles about CompuServe and competing information services AOL and Prodigy. When the *New York Times* reported in 1994 on the *San Jose Mercury News*'s online edition on AOL, for instance, the article included a subhead reading "A Media Convergence" and said the newspaper's executives "were convinced that technological changes were bringing all forms of media together."[7] This vision of the future seemed to accelerate in the mid-1990s, when the World Wide Web emerged into the popular consciousness and seemed to be the "one grand system" Pool had predicted. By the time AOL and Time Warner announced their landmark merger in early 2000, the word *convergence* had become commonplace in connection with electronic content delivery. In the meantime, though, another development in the news business had set the stage for yet another meaning of the word.

The new development was a decision by newspaper companies to partner with television news stations instead of just competing with them. These partnerships took different forms. In 1993, in Chicago, the Tribune Company launched CLTV, a twenty-four-hour local cable channel that used journalists and content from the *Chicago Tribune*. In other markets, newspapers paired off with local broadcast stations to cross-promote each other's content—in a negotiated partnership with no common ownership. The typical cross-promotional partnership included such elements as a "tomorrow's newspaper headlines" segment on TV, and a tie-in with the TV station's meteorologist on the paper's weather page. Finally, a handful of companies that owned both papers and TV stations in the same market—whose joint ownership predated federal prohibitions of the practice—became more aggressive about cross-promotion and content sharing. The most striking example came in Tampa, where owner Media General constructed a new building to house the news operations of the *Tampa Tribune*, WFLA-TV, and the newspaper's Web site, Tampa Bay Online.

Initially, newspaper-TV relationships weren't labeled as instances of convergence. Within the Tribune Company, which established cable news channels in both Chicago and Orlando, the more common term was *synergy*.[8] And most examples of newspaper/TV collaboration really didn't resemble the traditional ideas underlying technological convergence. Newspaper and TV journalists were not collaborating on content, and they were reaching audiences using traditional print and broadcast technology. But in trying to sell reluctant staff members on the merits of print-TV partnerships, advocates were able to evoke the future by using the word "convergence" as shorthand.

By this time, "convergence" was being used so frequently, in so many different contexts, that it had largely lost its value in focusing discussions about journalism and the news media. But at the same time, the popularity of the word was evidence of its power, especially among advocates for change in newsrooms, media companies, and journalism schools. In a single word, these change agents found a label that conveyed a sense that the world was being transformed and a vision for confronting the future. But the inescapable problem remains: This single word has many different meanings. In the remainder of this chapter, I'll try to explore their significance—in 2002 and in the future—for media companies, readers and viewers, journalism and journalists. First, I'll address the original, technology-rooted definitions, then the ways the term has been applied in media organizations.

Convergence in Media Technology

Pool's idea of a "convergence of modes" conjured up an image of a future where all content would be stored digitally, delivered over a network, and accessed through electronic devices. When the World Wide Web appeared on the scene, some thought that this future was near. But a few years later, it seems clear that it is, at best, quite some time away. In 2002, communications technology looks like this:

- *Content creation:* Most informational content (and some entertainment content) is created using computers and stored digitally. It is often stored in formats, however, that make it difficult to use for more than one purpose. A newspaper article formatted and stored with the layout program QuarkXpress cannot easily be translated into a Web page. A Web page formatted for delivery to a computer must be converted to be readable on an Internet-connected cellular phone. A digital videotape used for broadcast news must be "ingested" into a computer—and, because of bandwidth limitations, compressed—before it can be streamed over the Internet.
- *Content distribution:* More and more people (in the United States, more than half of all households and a greater proportion of businesses and schools) have access to the Internet. And thanks to the World Wide Web, companies that create content are able to distribute it digitally. But while the Internet has changed people's patterns of media usage, it is far from dominant as a distribution channel. As of 2001, according to investment banker and consultant Veronis Suhler Stevenson, the average American spent 149 hours per year (about 24 minutes per day) using the Internet, an amount of time comparable to that spent with newspapers and more than that spent with consumer magazines. But it was less than one-tenth the time spent with television and about one-sixth the time spent with radio.[9] In part, that is because more than one-quarter of Americans don't use the Internet;[10] in part, it is because dial-up connections—relied upon by almost nine in ten home Internet users[11]—are frustratingly slow.
- *Content consumption:* Other than the computer, most consumers do not use digital display devices to view information or entertainment. Although the U.S. Federal Communications Commission (FCC) has set a 2006 deadline for conversion to digital television, most TV viewers rely on analog programming delivered to traditional sets. Cable television companies are the leading providers of high-speed Internet access, but their Ethernet cables connect to computers, not to TV sets or set-top boxes. And while there was a flurry of interest in 2000 and

2001 in e-book readers as a way to distribute digital books, they failed in the marketplace.

In short, the era of complete technology convergence is not yet upon us. But we can see its outlines taking shape. The next decade will see a number of key technology developments:

- The creation and widespread use of digital *content management systems* within media companies, which will store content in digital formats—such as Extensible Markup Language (XML)—that allow it to be delivered relatively easily to different platforms.
- The proliferation of *wireless Internet access*, either through cellular telephone systems or through more localized wireless networks that, in turn, connect to the wired Internet.
- The *transformation of television* as TV sets take on more and more of the attributes of computers. For technology convergence to enter the living room, TVs will need access to the Internet, the capacity to receive and store digital content, and the ability for viewers to interact with content (and advertising) on the screen. All these technology changes are starting to happen. Internet access is available through cable Internet services. Digital broadcasting has already begun. Products such as TiVo and writeable digital video disc (DVD) drives allow the storage and playback of digital video. Cable and satellite TV providers have also begun—especially in Europe—to roll out interactive television services.
- A new generation of *portable devices* that come closer to replicating the advantages of paper will be lighter and will have longer battery life, improved screen resolution, and better viewability in bright light. These devices may be the descendants of today's laptop computers, personal digital assistants, or cellular phones, or they may emerge from laboratories and businesses trying to develop the technology for products with names like "electronic ink" or "e-paper."

For complete digital convergence to become a reality, then, we'll need to see technological changes in every stage of the information infrastructure: from content creation to distribution to consumption. But before that day comes—even if it never arrives—convergence in other forms is affecting the companies that produce information and entertainment, and the people who work in them.

Convergence in Media Organizations

Consider the media company as a hierarchical organization. At the top is a parent corporation, which in turn owns individual media properties. Within each company are the organizational structures, workflow processes, and tactics that reflect the way the enterprise operates. At the bottom are the individual employees who create the content—among them, the journalists. As of 2002, the word *convergence* can be applied to all of these: the companies, their operations, and the way their employees do their jobs. Using this framework, we can identify at least five different meanings of "convergence" and discuss their implications for different layers of a media company.

Convergence: Ownership

At the highest level of today's media conglomerates, "convergence" means the ownership of multiple content or distribution channels. In the global economy, Viacom, Disney, Vivendi Universal, and AOL Time Warner are frequently used as examples of what has been called ownership convergence.[12] In the United States, Tribune Company, Belo Corporation, and Hearst Corporation are examples of companies—smaller, but still large—that own newspaper, television, Internet, and other media properties. The merger between AOL and Time Warner in 2000 marked a particularly significant event in the annals of ownership convergence. It was, the *Financial Times* of London reported, "the clearest indication to date that the much anticipated coming together of programming of all kinds and the means to deliver it is becoming a reality."[13]

In the eyes of many media critics, ownership convergence raises significant concerns. In 1983, in the first edition of *The Media Monopoly*, journalist and academic Ben H. Bagdikian predicted that "a handful of corporations would control most of what the average American reads, hears and sees." And he worried that the concentration of ownership and control of content by companies with an interest in preserving the status quo would stifle the diversity of voices necessary to produce an accurate "picture of reality" in news coverage.[14] In the United States, long before technology convergence was considered, concern about the concentration of media ownership resulted in FCC rules that limited the cross-ownership of newspapers, television stations, and cable systems. In 2002, the FCC announced the start of a comprehensive review of these restrictions, some of which dated to the 1940s. The restriction that attracted the most attention was a 1975 rule that barred newspapers from owning broadcast stations in the same market (except for cities, like Chicago and Tampa, where joint ownership was already in place). In 2000, with the acquisition of Times Mirror, executives at Tribune Company placed a big bet that this restriction would be eliminated or modified. In acquiring Times Mirror, Tribune owned newspapers and broadcast TV stations in each of the top three U.S. markets: Los Angeles, New York, and Chicago. Their assumption was that by the time the New York and Los Angeles stations' licenses came up for FCC renewal in 2006, they would face no regulatory challenges. "Given the proliferation of media from 1975 to today, this rule doesn't hold water," a Tribune vice president told the *Washington Post*.[15]

It's worth pointing out that ownership convergence does not necessarily require shared editorial decision making or other kinds of collaboration across distribution platforms. In the months after their merger, AOL Time Warner executives were still struggling mightily to figure out how they could get their different media properties to work together more effectively. And Tribune Company owned both a newspaper and a broadcast station in Chicago for decades during which they operated with very little interaction. But as interest in "synergy" grew in corporate executive suites, Tribune increased the amount of cross-promotion and content sharing between print and TV. Tribune executives argued that ownership of television stations by newspapers could improve the quality of TV news. Given that newspapers employ many times the number of journalists that TV stations have, and given the pressure from Wall Street for publicly held companies to deliver ever-increasing profits, there is some merit to the idea of sharing resources. As Jack Fuller, president of Tribune's publishing subsidiary, put it, "Owning television, radio and newspapers in a single market is a way to lower costs, increase efficiencies and provide higher quality news in times of economic duress."[16]

The issues related to ownership convergence get increasingly complicated as more and more information is distributed electronically. Traditionally, newspaper publishers had control over the distribution system for their content, and broadcasters owned (at least for the duration of an FCC license) both content and a broadcast frequency. But as new distribution systems came into being—online services, Internet service providers, cable systems—critics of ownership concentration found a new area for concern. In the new world of electronic distribution, the companies with the closest connection to consumers were not the content creators, but the companies that provided the communications technologies. This was a primary area of focus for Pool in *Technologies of Freedom*, particularly because it was unclear what type of government regulation would apply to these new distribution channels. Would electronically distributed speech be subject to the First Amendment, or would it be subject to the kinds of restrictions imposed on TV and radio as a result of their government licenses?

For those worried about the potential power of the new electronic distributors, a noteworthy event occurred on May 1, 2000. Time Warner Cable, unwilling to meet the Disney Corporation's terms for the right to carry Disney stations, simply pulled the ABC television network from its cable systems. Assailed by outraged consumers, politicians, and federal regulators, Time Warner relented the next day and restored ABC programming. But the incident illustrated a shift in the balance of power between programmers and distributors. "This is a sign of things to come with the AOL/Time Warner merger," a Disney executive told the *Los Angeles Times*.[17]

Convergence: Tactics

Perhaps because of the rise of the World Wide Web, perhaps because of the inexorable fragmentation of the mass media, traditional media companies in the late 1990s began engaging in a variety of activities that I'll call tactical convergence. These activities fell into three general areas: content, marketing, and revenue enhancement.

Just as ownership convergence does not necessarily imply collaboration on tactics, tactical convergence does not require common ownership. In the content and marketing arenas, the most common model was a partnership between a TV station and a newspaper, in most cases (but not all) under separate ownership. In most markets, the primary motivation for—and initial results of—these partnerships seemed to be promotional. The partners assumed that cross-promotion would drive newspaper readers to watch the local news and TV viewers to read the newspaper. The TV viewer would hear about a story in tomorrow's paper. The reader of the newspaper's weather page would come to rely on the station's meteorologists for up-to-date forecasts. In more aggressive markets, especially those where a single company owned both outlets, a common result was "talkbacks"—appearances by print journalists on the air to talk about stories they were covering.

Journalists involved in tactical convergence had a wide range of reactions. At newspapers, some embraced the partnership as a way to increase their audiences. And on the TV side, the newsroom often saw the advantage of adding many more "feet on the street" to report the news (e.g., 225 reporters at the *San Francisco Chronicle*, 16 at the local CBS broadcast affiliate, KPIX[18]). But there were also significant challenges. The first person to raise concerns was often the newspaper's TV critic, who wondered how he or she could be perceived as objective in covering all the stations in the market when the newspaper had a high-profile relationship with just one. There were also cultural stereotypes to overcome. Newspaper staff members saw the TV journalists as shallow and more interested in image than substance. The TV staff saw the print reporters as rumpled, hostile, and unappreciative of the challenges involved in putting together a good broadcast news piece. In both newsrooms, there was reluctance to allow the other to publish an enterprise story first.

Despite these concerns, advocates of tactical convergence argued that it benefited journalists and the public. Forrest Carr, news director of WFLA-TV, one of the partners in Media General convergence effort in Tampa, described seven levels of interaction between his station and the *Tampa Tribune*.[19] Because of their joint ownership and Media General's emphasis on cooperation between the two, his examples of collaboration are more numerous and more pervasive than would be found in partnerships between separately owned companies. But he provided a good list of the kinds of collaboration that are possible in print-TV tactical convergence:

- *Daily tips and information.* Communication occurs through scheduled meetings and spur-of-the-moment conversations between assignment editors from the two newsrooms throughout the day.
- *Spot news.* Newspaper reporters enable the TV station to provide more detail and depth in their live reports from the scene of a news event.
- *Photography.* The TV station's camera crews carry still cameras; newspaper photographers carry digital video cameras. "There's no sense in . . . sending separate photographers to a ribbon-cutting, if all the paper needs is a single shot and all the TV station needs is 25 seconds of video," Carr wrote.
- *Enterprise reporting.* A TV reporter's investigation into bridge corrosion was given to the newspaper to publish first, resulting in a 25 percent ratings increase in that evening's news.
- *Franchises.* This TV term for regularly aired features is now applied to regularly scheduled, cross-platform content. Examples include a newspaper religion reporter who appears on air on the same day her print feature is published, or a TV consumer reporter who writes a column for the paper.

- *Events.* WFLA didn't send a reporter to cover the Winter Olympics, but the Tribune did. So the newspaper's reporter appeared on air nightly with Olympics highlights.
- *Public service.* The paper and TV station cosponsored a town hall meeting about the responsiveness of local news and was planning to collaborate on political campaign coverage.

One noteworthy aspect of Carr's article was its emphasis on TV-newspaper collaboration and the relatively little mention made of the partners' affiliated Web site, TBO.com. In the wake of the collapse of the dot-com boom, this was not uncommon at tactically converged news operations. In some cases, this was because TV stations and newspapers maintained separate Web sites that considered themselves competitors in the online market. But even in markets like Tampa, where TBO.com collaborated with both TV and newspaper, Web sites were increasingly viewed as junior partners—a distribution platform for content created by the other two organizations, rather than a third peer. With staffs pared down because of a decline in online advertising and increased emphasis on balancing revenues with expenses, online operations were ill equipped to supply original content or add unique-to-the-medium elements.

One other area of tactical convergence deserves mention: Efforts to sell advertising packages encompassing multiple platforms. Participants in efforts to sell multiplatform advertising found even greater challenges than their counterparts involved in multiplatform news coverage. For one thing, advertising sales representatives tend to specialize by medium—a TV sales rep understands the strengths of TV and how to sell them to an advertiser—and also to "sell against" the other types of media. In addition, sales representatives are innately possessive of their client relationships and reluctant to collaborate with other salespeople. In addition, in markets with common ownership, there was a realistic concern that joint ad sales would produce less revenue, rather than more, because the advertiser would insist on a lower price for a multiplatform package than for separate ad buys in different media. In the case of online advertising, there were other problems. Print and TV ad sales staffs didn't understand online advertising or how to sell it, and traditional print and TV advertisers were slow to develop interest in online ads. So it took longer to complete an online ad sale. Since online ads cost a fraction of TV and newspaper advertising, many ad sales representatives didn't see a reason to commit much energy to online advertising. Still, big companies were betting that they could generate new revenue by selling ad packages across TV, newspaper, and online. Tribune

Company announced a 2001 goal of $16 million in ad sales by Tribune Media Net, its cross-media team, and in 2002 announced a major cross-platform deal with Target Corporation.[20]

Convergence: Structure

Ownership and tactical convergence don't necessarily require significant changes in organizational structure or the way employees do their jobs. In markets where TV, newspaper or online operations collaborate, it's clear that most staff members focus their energies on the primary medium they work for. But the more aggressive the goals for convergence, the more likely it is that job descriptions and organizational structures will change. Here are a few examples of what could be called structural convergence:

- The *Orlando Sentinel*, which launched a twenty-four-hour local cable news channel in partnership with Time Warner Cable, created a staff of multimedia editors whose job it is to do whatever is necessary to get newspaper content—and print reporters—on the air. The editors, most of whom come from broadcast backgrounds, coordinate between the two newsrooms, arrange talkbacks for print reporters, and produce original TV programming, such as a weekly high school sports show.
- Startribune.com, the Web site of the *Minneapolis Star-Tribune*, hired a TV photographer and producer to serve as an online multimedia reporter. She covers news events, shoots video, takes still photographs, and creates multimedia presentations for the Web site.
- The *Indianapolis Star* and WTHR-TV, the local NBC affiliate, agreed to share the cost of a salary for a director of news partnerships whose job was to foster content collaboration between print and TV. The director, Jon Schwantes, had previously been a reporter and editor at the *Star*.
- In Sarasota, where the *Herald-Tribune* started its own local cable news station in 1995, the newspaper's executive editor also was made responsible for overseeing the TV operation. In 2002, the company created a new position, general manager for electronic media, with operational responsibility for news on the TV station and the paper's Web site. The person who got the job, Lou Ferrara, was formerly city editor for the newspaper.

In a few newsrooms, structural changes can be seen with the naked eye. In Chicago, for instance, a TV stage sits in the middle of the *Tribune* newsroom. In Orlando and Tampa, new assignment desks were built to foster interaction between print, TV, and online editors.

Convergence: Information Gathering

Among journalists discussing convergence, no topic generates more heated debate than whether it is likely or desirable for individual journalists to report a story using multiple media tools. The subject is particularly controversial for print reporters, who don't see themselves carrying video cameras or audio recorders around as they do their jobs. At Columbia University, engineers have developed a mobile journalist workstation, which straps on to a reporter's back and enables him or her to return from a news event with multiple types of content. When I've shown pictures of it to journalists, there are always snickers because it looks so bulky and cumbersome—though obviously, technology advances will ultimately allow such a device to be much smaller and lighter.

In 2002, the idea of the backpack journalist generated strikingly different outlooks in a point-counterpoint pair of articles on the Online Journalism Review Web site. Jane Ellen Stevens, who has worked as a newspaper reporter and video producer for television and online delivery, was proud to call herself a backpack journalist. "These days, can you imagine hiring a reporter who doesn't know how to use a computer?" she wrote. "In 10 years, you won't grok hiring a reporter who can't slide across media, either."[21] (The word *grok* is a literary reference that means, roughly, "consider," "understand," or "contemplate.") As an example, she cites Preston Mendenhall of MSNBC.com, who spent two weeks traveling Afghanistan and sent back written articles, still photos, audio, and video—his work was presented on air and on the Web. The companion article, "The Backpack Journalist Is a 'Mush of Mediocrity,'"[22] quotes journalists who claim that examples like Mendenhall's will be—and should be—rare.

However rare these multimedia reporters may be in the future, it's clear that they increase in number as time goes on. And even if most journalists aren't expected to produce content for every conceivable delivery platform, some are already being asked to gather information in multiple formats. At twenty-four-hour local cable news operations, for instance, it is common to expect journalists to write stories, shoot video, and edit it themselves. This is a sharp contrast from traditional broadcast news stations, where reporting, news photography, and video editing are discrete professions. Or consider the *Topeka Capital-Journal*, where newspaper reporters are routinely expected to tape-record their interviews and bring them back to the newsroom for presentation on the paper's Web site, or the *Orlando Sentinel*, where newspaper photographers also frequently shoot video for their local cable news partner.

Convergence: Presentation (Storytelling)

With every new medium has come a series of conventions for presenting information—or, to put it another way, for telling stories. These conventions take time to evolve. The early TV news broadcasters simply sat behind a desk and read articles that differed little, if at all, from those written for newspapers. But ultimately, TV journalists figured out how to take advantage of their medium's unique capabilities. As we move toward Pool's convergence of modes, it is reasonable to expect that new forms of storytelling will emerge for the three new digital presentation platforms: desktop computers, portable devices, and interactive TV. These platforms have some unique capabilities:

- They have the potential of unlimited space and time, so journalists are no longer constrained by column inches, or pages, or minutes on the air.
- They can allow immediate publishing, as news breaks, independent of press deadlines or programming schedules.
- They can facilitate communication with and among readers and viewers, giving the audience members an unprecedented ability to react to or shape the content, or even to supply content themselves.
- They can deliver content in multiple media formats: text, photographs, graphics, audio, video, and animation.
- Perhaps most importantly, they are interactive, meaning that the user has great control over the content, deciding what to view, in what order, and when to move on to something else.

On the Web, the first new "converged" presentation medium, journalists quickly began to experiment with new storytelling forms. Chicagotribune.com created an innovative presentation that allowed users to get a sense of what it was like to stand in the batter's box facing a hard-throwing young pitcher. The *New York Times* created packages incorporating text, audio, video, and 360-degree panoramic photos. A small paper in Washington State, where the local government was trying to make decisions on waterfront development, created an interactive game allowing citizens to choose the types of developments they wanted. And many news sites experimented with a powerful new storytelling tool: an animated slide show of still photographs, sometimes accompanied by music or recorded sounds.

In the early days of Web journalism, innovative storytelling did not attract much of an audience. Perhaps this was because most of the audience, using slow dial-up connections, couldn't appreciate these stories. Or perhaps it was because Web users were so task-oriented that they didn't want to be distracted by these elaborate interactive presentations. But as time went by and bandwidth increased, news sites began developing sizable audiences for their multimedia storytelling. In March 2002, when part of a scaffold blew off a Chicago skyscraper, killing three people, chicagotribune.com's photo essay got more traffic than the text story about the event. After the terrorist attacks of September 11, 2001, multimedia features were among the most popular content on news Web sites.

As the audience for examples of presentation convergence grows, news organizations will have to figure out how to produce it regularly at reasonable cost. One model might be to create more backpack journalists who can do everything necessary to create an interactive multimedia news story. I think the more likely outcome is greater collaboration among journalist teams. A multimedia producer, using content gathered by reporters, photographers, videographers, and graphic artists, will produce packages for the new digital media.

Convergence: Implications

Technological convergence continues to move forward. Even more significantly, more and more of the media audience is comfortable in a world where information streams in through multiple channels. But as of 2002, convergence in media organizations is clearly most prevalent in its ownership and tactical forms. In selected companies, some new kinds of jobs have been created and new ways of doing things adopted—examples of what I've called structural convergence. Information gathering and storytelling convergence, however, remain relatively uncommon—and, as a result, most journalists' jobs have changed little. The progression from ownership and tactical convergence to information gathering and storytelling convergence will be slow. Ownership and tactical convergence can be implemented without transforming the culture of a media company, though it will probably be more successful the more the culture changes. For information gathering and storytelling convergence to become commonplace, many new kinds of jobs will need to be created, and many existing jobs will require new skills. It may, then, be a good thing for today's media companies that complete technological convergence is not yet upon us. And for journalists who are averse to change, who like journalism the way they have traditionally done it, the relative lack of change is comforting. But it seems safe to predict that greater changes—in the form of new jobs, new job requirements, and new opportunities—lie ahead. For individual journalists, the companies that employ them, and the colleges and universities preparing the next generation, the present and future of convergence have significant implications. At a minimum, all journalists will need to develop a basic understanding of the unique capabilities of the different communications media. Increasingly, their employers are going to deliver content to multiple platforms or collaborate with other companies to do so. No longer can journalists assume that just because they work in one medium (say, a print newspaper), they don't need to worry about how their story should be presented in another (on TV or the Web). No longer can journalism school faculty assume that they can turn out graduates who understand only one set of communications tools. On the other hand, we are not necessarily moving into an era when a single journalist needs to do it all—report, write, take pictures, shoot and edit video, and present their stories on the Web. There will always be a need for specialists who do one thing particularly well. But in the converged media organizations of the future, the journalists who best understand the unique capabilities of multiple media will be the ones who are most successful, drive the greatest innovations, and become the leaders of tomorrow.

Notes

1. "Convergence, *n*," *Oxford English Dictionary*, J. A. Simpson and E. S. C. Weiner, eds. (Oxford: Clarendon Press, 1989).

2. David E. Carlson, "David Carlson's Online Timeline," at http://iml.jou.ufl .edu/carlson/professional/new_media/timeline.htm, accessed September 12, 2002.

3. Ithiel de Sola Pool, *Technologies of Freedom* (Cambridge, Mass.: The Belknap Press of Harvard University Press, 1983), 27–28.

4. Pool, 1–2.

5. Stewart Brand, *The Media Lab: Inventing the Future at MIT* (New York: Viking Penguin, 1987), 10–11.

6. Reproduced in Paul J. H. Schoemaker, *Profiting from Uncertainty: Strategies for Succeeding No Matter What the Future Brings* (New York: The Free Press, 2002), 83–84.

7. William Glaberson, "In San Jose, Knight-Ridder Tests a Newspaper Frontier," *New York Times*, February 7, 1994, D1.

8. Ken Auletta, "Synergy City: Chicago's Tribune Co. Is Revolutionizing How It Does Business—But at What Cost to Its Newspapers?" *American Journalism Review*, May 1998, 18.

9. Veronis Suhler Stevenson, *Communications Industry Forecast* (New York: Veronis Suhler Stevenson, 2001), 42–43.

10. UCLA Center for Communication Policy, The UCLA Internet Report 2001—"Surveying the Digital Future" (Los Angeles: UCLA Center for Communication Policy, 2001), 17.

11. UCLA Internet Report 2001, 25.

12. Everette E. Dennis and John V. Pavlik, "The Coming of Convergence and Its Consequences," in *Demystifying Media Technology*, John V. Pavlik and Everette E. Dennis, eds. (Mountain View, Calif.: Mayfield Publishing Co., 1993), 1–3.

13. Alan Cane, "Convergence Is the Watchword," *The Financial Times*, March 15, 2000, 1.

14. Ben H. Bagdikian, *The Media Monopoly* (Boston: Beacon Press, 1990), x–xvi.

15. Howard Kurtz, "Tribune Goes Through Channels to Keep TV," *The Washington Post*, March 20, 2000, C1.

16. Kathy Bergen, "Survival Top Story for Independent Papers," *Chicago Tribune*, September 11, 2002, Section 3, 1.

17. Sallie Hofmeister, "Cable TV Dispute Cuts off ABC for Millions of Viewers," *Los Angeles Times*, May 2, 2000, 1.

18. Deborah Potter, "The Body Count," *American Journalism Review*, July/August 2002, 60.

19. Forrest Carr, "The Tampa Model of Convergence," at www.poynter.org/centerpiece/050102_2.htm, accessed May 6, 2002.

20. Lucia Moses, "Tribune Media Net's fast start," *Editor & Publisher Online*, August 7, 2000, at http://editorandpublisher.com/editorandpublisher/search/article_display.jsp?vnu_content_id=1124588, accessed September 13, 2002.

21. Jane Stevens, "Backpack Journalism Is Here to Stay," *Online Journalism Review*, April 2, 2002, at www.ojr.org/ojr/workplace/1017771575.php, accessed September 13, 2002.

22. Martha Stone, "Backpack Journalism Is a 'Mush of Mediocrity'," *Online Journalism Review*, April 2, 2002, at www.ojr.org/ojr/workplace/1017771634.php, accessed September 13, 2002.

4

New Technology and News Flows: Journalism and Crisis Coverage

John V. Pavlik

D URING A CRISIS, JOURNALISM PLAYS arguably its most important role in society. News serves as a lifeline connecting the public and policy makers to information about critical events as they unfold. But how does news flow during crises that may knock out regular means of news dissemination and communication? Such was the case during the September 11, 2001, attack on the World Trade Center in New York City as more than one hundred television and radio transmission antennae were destroyed when the One World Trade Center tower collapsed. In addition, damaged telecommunications equipment and massive call volume hampered the telephone system as journalists and the public alike were unable to make voice calls.

This chapter examines the role of new and emerging information technologies in the distribution of news and information during moments of crisis, with a particular focus on the United States and North America since September 11, 2001. Among the technologies examined are satellite communications and remote-sensing wireless Internet communications, and mobile information-acquisition devices.

Various information technologies play an increasingly important role in the dissemination of news and information during a crisis. These technologies are increasingly digital, wireless, and mobile and provide Internet connectivity. They tend to serve important functions in both news gathering and communication among journalists producing news coverage, as well as giving audiences access to news during a crisis. These technologies play an increasingly significant role in facilitating audience communication with journalists, sometimes providing an opportunity for members of the public to contribute their own reports to the flow of news and information, thereby expanding news coverage, but also raising the potential for misinformation (since members of the public may not have the same professional training, experience, or standards of fact checking used by most journalists and news organizations).

Wireless Mobile Communications

Among the most important tools for news and information distribution during times of crisis are wireless mobile communications. At this point, it is worth briefly defining the nature of a crisis. For the purposes of this paper, a crisis is defined as a crucial or decisive point or situation, often characterized by political, social, or economic instability, typically triggered by a sudden change or traumatic event. Clearly, the terrorist attacks on the World Trade Center in New York City and the Pentagon in Washington, D.C., on September 11, 2001 (and the hours and days, perhaps even weeks, immediately following the attack) fit this definition.

As the tragic events of September 11, 2001, unfolded, mobile wireless communications repeatedly demonstrated their value. In fact, at least one passenger on United Airlines flight 93, one of the hijacked flights, was able to use a cellular phone to establish voice communications with persons on the ground during the ill-fated flight. These cellular communications not only played a role in helping the passengers organize their resistance to the hijackers, but also helped inform those on the ground as to what transpired on the plane. "A GTE AirFone operator, speaking with one of the passengers on board, reported that the passenger had told her that the passengers had voted on whether to obey the hijackers, or challenge them, and that the vote had been to attempt to regain control of the plane" (Chris Kilroy, AirDisaster.com, www.airdisaster.com/special/special-0911.shtml).

With much telecommunications equipment in New York suffering extensive damage and massive voice call volume hampering the landline-based telephone system, mobile telephony and wireless e-mail (e.g., Research in Motion's Blackberry) proved to be among the most reliable means of communication. In fact, e-mail in general was often more accessible than voice telephony.

A growing variety of wireless communication technologies that are emerging may prove vital to news flows during future times of crisis. An increasingly popular wireless local-area networking technology known colloquially as Wi-Fi and technically as IEEE 802.11b is being installed throughout New York, San Francisco, and various other communities and university campuses around the United States and the world. Wi-Fi is broadband (high-speed, with transmission rates up to 11 Mbps, comparable to landline Ethernet connections and faster than many commercial landline Internet services such as digital subscriber lines and cable modems). Further, Bluetooth, another high-speed wireless local-area network, is being deployed in a variety of digital devices. In Japan, the i-mode wireless Internet phone in Tokyo from NTT-DoCoMo is proving popular among the public and media professionals alike. As a result, journalists and the public will likely use mobile wireless technologies increasingly during times of crisis, when mobility is at a premium.

Miniature Digital Cameras

Of critical importance to news gathering in times of crisis is the emergence of quality digital cameras, both still image and motion video. Increasingly, inexpensive digital cameras are capable of capturing megapixel images. (Megapixel means the image contains at least one million pixels, a pixel being the smallest element, or point of light, in a digital image.) Sony introduced in the summer of 2001 a digital camera for less than $1,000 that can capture 5.1-megapixel images. This same camera can shoot and capture MPEG video (320×240 pixels of resolution at sixteen frames per second). This is less than broadcast quality (640×480 or five hundred horizontal lines of resolution, full screen at thirty frames per second). But it is more than adequate for exclusive images and video during times of crisis.

Consider the case of Andrea Booher, a photographer hired by the Federal Emergency Management Agency (FEMA) to shoot photos of

"The Pile," as the remains of the multibuilding complex at Ground Zero have been dubbed:

> As one of two official photographers for FEMA at Ground Zero, Booher had unfettered access to the disaster site. Her job was to photograph the rescue and cleanup effort. FEMA would make some of her pictures available to the media on its Web site, but the agency also was creating a pictorial history of the event for later reference. (Mobile Computing Online, February 2002, p.1, at www.mobilecomputing.com/fullarchives.cgi?195)

This could not have been done without a digital camera, in this case a Nikon D1X. "It saved me," Booher says. "There is no way I could have kept up with film." Processing was too problematic, the demand for images was high right after the disaster, and labs were too busy or closed (Mobile Computing Online, February 2002; no longer available on the Web).

Importantly, digital images and video can be captured and stored either at full resolution or in compact form for easy distribution as an e-mail attachment, even via wireless devices where bandwidth may be limited. Further, a growing number of digital cameras come equipped with night vision, or infrared image capture technology, making it possible to shoot at night or under other low-light conditions. Technologies such as holographic autofocus and a full range of manual settings and built-in flash also aid the photographer or videographer in need of flexibility during times of crisis.

During the 1990s war in the Balkans, Radio-B92, an independent station that stayed on the air via the Internet, sent a reporter behind the battle lines to capture images of the war-torn scene. But, military officials would not allow journalists with cameras behind the lines. So, with the help of a compact digital camera that he could roll up in his shirt sleeve like a pack of cigarettes, the reporter was able to sneak his camera past the military and capture images of the war—the journalist felt that documenting the situation was worth the considerable risk he took in smuggling the camera past the officials. After returning back across the lines, the photos were published on the Radio-B92 Web site, reported Serbian journalist Drazen Pantic (personal interview, February 7, 2001).

Mini digital video cameras are also increasingly sophisticated and compact, capturing full broadcast-quality video, as well as megapixel still images. Cameras such as the Canon XL1 or the Sony VX1000 both produce video with five-hundred-plus lines of resolution at thirty frames per second. Moreover, these cameras are fully digital and are equipped with FireWire (also known as IEEE-1394) to transfer the video in real time to a laptop computer for editing with video editing software such as Adobe Premier or Apple Final Cut Pro.

Satellite phones are also playing an increasingly important role in news and information flows during times of crisis. During the war in Afghanistan (fall 2001 to spring 2002), journalists relied extensively on satellite phones to report from battlefields or other remote locations where no landlines existed for standard voice communications. They also used satellite phones to upload video (typically less than broadcast quality) of scenes of the conflict. By combining digital camera and microphone, laptop computer, and satellite phone, journalists can report anytime from nearly anywhere on the planet.

Satellite phones have been used in journalism before, as early as the Gulf War in 1990. But it was the downing of a U.S. spy plane on the Chinese island of Hainan in the spring of 2001 that stimulated heightened interest among news organizations to use satellite phones during times of crisis or breaking news from remote locations. As it turned out, only CNN had a satellite phone to transmit live video of the crew leaving Hainan Island, while competing news organizations had to travel to a satellite base station minutes away; thus, CNN beat them in providing video of the breaking story by roughly thirty minutes. Consequently, many other major news organizations acquired satellite phones for use in future crisis situations.

Digital audio capture offers journalists another important advantage during times of crisis, or even when simply on deadline. Digital audio can be fed directly into a laptop computer equipped with speech recognition software such as Dragon Naturally Speaking or IBM's Via Voice to obtain an instantaneous transcript. These technologies now work with natural speech (i.e., discreet speech, or speech in which the speaker pauses briefly between each word, is no longer needed), without having to train the speaker or customize the system to each speaker. Although reliability is less than 90 percent when working without customizing the system to the speaker, a real-time transcript can greatly aid the reporter in searching through a large audio file for just the right actuality or sound bite, especially when on deadline. Moreover, the transcript can be cleaned up by the reporter for subsequent posting to the Web or for incorporation into a written news story.

Crisis reporting is also being aided greatly by a variety of other emerging sensors. Among these sensors are cameras with unusually large fields of view (e.g., 360-degree cameras), high-resolution satellite imaging, face-recognition systems, and ground-penetrating radar.

Omnidirectional, or 360-degree, cameras have been used for reporting since the 1990s, initially for feature reporting mostly, but increasingly for hard news. In the 1999 slaying of Amadou Diallo, an unarmed black man in the Bronx, New York, shot and killed by four plainclothes white police officers, reporters in the author's course, "The News Laboratory," used a 360-degree camera invented at Columbia University by computer science professor Shree Nayar to document the slay site. The 360-degree images, which permitted the viewer to navigate about the scene and examine the context of the slaying, were published by the award-winning online news service, APB News.

Various 360-degree imaging systems have been developed and are commercially available today. Among the most used by news organizations are Remote Reality and Ipix. Two others, Fullview and iMove, have systems that can produce full-motion broadcast-quality 360-degree video. Their use by news organizations is likely to increase as digital and interactive television emerge in the next decade, first in sports and feature reporting, and later in crisis situations, as the technology becomes more portable and familiar.

On the horizon is the Cyclops, a system Professor Nayar has developed in his Computer Automated Vision Environment (CAVE) Lab. The Cyclops is a small (six-inch) working prototype of a 360-degree video and audio capture system for use in crisis situations. A microphone mounted on a gyroscope maintains the correct orientation all of which fits inside a six-inch Plexiglas sphere. The system includes a wireless transmitter, enabling remote viewing. One version of the Cyclops can be tossed or rolled into a room or other location, perhaps a burning building or hostage situation, and enables firefighters or authorities to observe and assess a situation before risking human lives in a possible rescue operation. Another version of the Cyclops has a motorized system of wheels that enables a remote operator to maneuver the sys-

FIGURE 4.1
A 360 photo of Dealey Plaza, the site where President John F. Kennedy was assassinated. On the upper right is the grass knoll where a second gunman might have crouched; to the lower left is the Texas School Book Depository building where Lee Harvey Oswald fired the fatal shot. The photo was taken using a RemoteReality OneShot360 Immersive Imaging System. Imagine if Abraham Zapruder had held a 360 camera. (Photo by Raghu Menon.)

tem as desired, around tables or chair or other obstacles. There's no reason why journalists could not make use of a future Cyclops for news-gathering purposes.

Chris Csikszentmihalyi, a scientist at the MIT Media Lab, envisions a robo-reporter that might be used in conflict situations such as those in Afghanistan to help reporters gather the news without unnecessarily risking their lives. If such a remotely controlled news-gathering robot could be developed without sacrificing news quality, it could prove extremely valuable; 2001 saw the death of thirty-seven journalists in the line of duty worldwide (www.cpj.org; www.washtimes.com/world/20020328-12813952.htm), including Daniel Pearl, a reporter for the *Wall Street Journal* captured and executed by terrorists in Afghanistan.

Remote-Sensing Satellite Imaging

When a U.S. spy plane was downed on the Lingshui Military Airfield, Hainan Island, in the South China Sea in April 2001, the world waited anxiously for a clear picture of just what was happening. Meeting that urgent need was a photo taken from 422 miles up in space. The photo was a 1-m resolution (i.e., objects as small as one meter in size can be discerned) color image taken by Space Imaging's IKONOS satellite at 10:12 A.M., local time, on April 4. The image pictured in figure 4.2 shows the U.S. Navy EP-3 "Aries II" sitting on a runway at Lingshui Military Airfield, Hainan Island, South China Sea. This image was published widely by the world's media, including newspapers, news magazines, television news programs, and news Web sites.

Remote-sensing satellite imaging has been around since at least the 1960s, when the U.S. military launched satellites to spy on the Soviet Union and weather forecasting began to rely on satellite imagery. With the end of the Cold War, the military has declassified much of the high-resolution satellite-imaging technology, largely to help spur economic development and commercial uses of high-resolution satellite imaging. Commercially available satellite imaging continues to improve in terms of the quality of the pictures, and as of this writing, satellite images of 0.6-m resolution are now commercially available. The military, it is suspected, has much higher-resolution imaging, but that information is classified and a well-kept secret.

Commercial uses of high-resolution satellite imaging are widespread, ranging from agriculture to archeology. Journalism applications have been fairly widespread since at least the Gulf War, when satellite images were used to help the public better understand developments in the conflict. Since then, news applications have ranged from military reporting to tracking refugee movements, deforestation, and many other stories.

In times of crisis, journalists are most likely to use satellite imaging to gain access to denied areas. Access to areas may be denied for a variety reasons. During natural or human-made disasters, environmental crises, or military conflicts, government authorities may not want reporters on the ground monitoring developments too closely for reasons of safety (of the reporters or military or civilian personnel) or to avoid disclosure of military strategies, interference with military actions, or

FIGURE 4.2
A U.S. Navy plane sitting on a runway at Lingshui Military Airfield, Hainan Island, South China Sea. (Satellite image by Space Imaging, www.spaceimaging.com.)

perhaps reporting on misdeeds or mistakes. During some crises, time may prevent reporters from getting to the scene quickly. Consequently, images taken from space may be especially useful in giving the public a visual understanding of developments. Images from space may also help place things into context better.

Space Imaging (www.spaceimaging.com) is a leading provider of satellite images used by the media. Figure 4.3 shows Space Imaging satellite images taken before and after the attack on the World Trade Center on September 11, 2001.

One of the troubling issues facing news media use of satellite images, especially during times of crisis, is "shutter control." Shutter control is an important First Amendment battleground. It refers to the fact that the government (i.e., the executive branch) has the authority to effectively close the shutter on the satellite cameras kept in space, either by government agencies or by private companies, at least those operated by U.S. companies. Satellites maintained by firms or agencies based outside the United States may not adhere to U.S. government requests to block the distribution of satellite images to commercial or other nongovernmental organizations, although most do so typically for political or economic reasons. The U.S. government has not yet employed formal shutter control, but during the weeks following the attack of September 11, 2001, the U.S. government employed an alternative satellite imaging censorship strategy: It spent millions of dollars (at least that is the amount suspected to have been spent—the government

FIGURE 4.3
The World Trade Center, before and after September 11, 2001. (Satellite image by Space Imaging, www.spaceimaging.com.)

has not yet revealed the exact amount publicly) to buy exclusive rights to satellite images involving possible terrorist activity, effectively denying the media, and thereby the public, access to satellite imagery during the most significant crisis faced by the United States in many years. Whether access to the images may have posed a threat to national security is not clear.

Two technologies increasingly employed commercially that have yet to become widely used in journalism are digital watermarks and the global positioning satellite (GPS) system. Yet, both of these technologies are likely to have an important impact on news reporting, especially during times of crisis. A digital watermark can take various forms, but essentially it is a form of a computerized stamp placed on a computerized image, video, audio file, or document. It can be visible or audible, or it can be invisible to the audience or user. It can be inserted at the point of acquisition (e.g., when an image or video or audio clip is captured) or during editing. A digital watermark is there largely to protect copyright. But digital watermarks can serve another important purpose, especially for news reporting during a crisis. Digital watermarks can help establish the veracity or authenticity of content, whatever its form.

One of the most powerful forms of watermarking technology involves the use of a GPS stamp. Because a GPS stamp records the exact latitude, longitude, altitude, and time (Greenwich Mean Time, or GMT) and is not easily modified, a reporter with a camera equipped with a GPS stamp could use it to document the exact location and time of the photo or video involved. In the case of a massacre, mass grave, or other crisis situation, verifying the location and time of an exclusive and potentially controversial image or video might be extremely valuable.

Technologies now in development present potential new tools for improved news reporting during crises. Among these technologies are dynamic ranging imaging, ground-penetrating radar, robotic sensors, and other intelligent sensors, such as cameras equipped with face-recognition software. Also on the horizon is augmented reality, a technology now being tested for both military and journalistic applications but already in use in medicine and manufacturing.

Dynamic-range imaging is an experimental digital photographic or videographic technology in which an image or video can be differentially exposed. In standard film or videography, an entire image or frame of video is exposed to light at the same level. As a result, if shooting into a dark window or doorway, it is impossible to see inside the darkened space without overexposing the surrounding objects. With dynamic-range imaging technology developed by Professor Nayar (www.cs.columbia.edu/CAVE), it is possible to solve this problem. Each pixel is exposed differentially. With the use of a digital "detail brush," the viewer can select different portions of an image or video, instantly change the exposure level, and see inside a darkened doorway, window, or cave. In times of crisis, such a technology could allow reporters to record scenes where uneven lighting might prevail.

Ground-penetrating radar has been developed that allows geophysicists, forensic scientists, and archeologists to "see" beneath the ground without digging. Objects of various types and sizes can be located. Future journalistic applications might include detecting or depicting a lost grave site or buried artifacts.

A variety of advances in robotics and sensors equipped with "intelligence" present potentially powerful opportunities for reporting, especially during times of crisis. Face-recognition technology, such as that developed by Visionics, now part of Identix (www.identix.com), is already in use by law-enforcement agencies around the world as a tool for identifying suspected criminals and terrorists, especially at points of controlled access to buildings or facilities, such as a metal detector in an airport. The technology is 90 percent reliable in recognizing and matching faces in a database of known criminals or suspected terrorists. Although its proponents say adequate privacy safeguards are in place (only matches are recorded), concerns remain (e.g., false matches are recorded as well, and it's not clear what happens to digital data that might be sent over a network to a remote server or hard drive).

Will news media begin to employ or have access to such face-recognition technology? There are no laws preventing it. The technology is so new that it is still far ahead of the legal system. (Just five years ago face-recognition technology existed only in experimental laboratories.) If a news cam equipped with face-recognition technology could spot known criminals at large in a community, what might be its value on the evening news? Imagine this headline: "Channel X's Smart News Cam Spotted America's Most-Wanted Felon Today on the Corner of A and B Streets." Or, what about a smart camera that automatically tracks celebrities?

Augmented Reality

Also on the horizon is a technology called augmented reality, which is a cousin of virtual reality. Instead of replacing a person's experience of reality with a computer-generated environment, augmented reality adds to a person's experience with the world around him or her. Many augmented-reality applications to date have focused on indoor applications, for instance, for manufacturing (e.g., permitting an welder to see precisely where to make a weld on an airplane wing) or medicine (e.g., aiding a physician in preparing and performing delicate surgery by seeing a three-dimensional overlay of a patient's internal anatomy before making an incision).

Through the use of various portable and wireless technologies, mobile augmented-reality systems can be developed for persons moving through the outside world. For example, equipped with a wearable computer, a see-through head-worn display, a GPS receiver (to determine the user's exact location), a head tracker (to determine the user's orientation within the environment), and a three-dimensional geographic information system for modeling the environment surrounding the user, a person can see and hear multimedia presentations embedded onto the real world in either translucent or opaque form. These multimedia presentations can be synchronized with or attached to objects, such as buildings, in the real world.

Through wireless Internet access, text, graphics, or multimedia can be streamed in real time to the user for visual display or audio playback. In this fashion, a journalist (or anyone else, for that matter) can travel down a city street and see a detailed three-dimensional map of where he or she is or see textual annotations placed on various objects or buildings providing historical or contemporary information. Military applications are also significant.

In collaboration with Columbia University computer science professor Steve Feiner and the Columbia Computer Graphics and User Interfaces Laboratory, I have explored the potential uses of mobile augmented-reality systems for journalism, developing a prototype system called the Mobile Journalist's Workstation (MJW). The MJW can be a useful tool for journalists in the field, especially in times of crisis when improved access to information and communications may be critical.

The MJW has been tested as both a news-gathering tool with conventional and omnidirectional video-gathering capabilities and a news-presentation system. As a news-presentation system, the MJW has been used to create a new form of documentary called a situated documentary, which enables news consumers to visit the site of past news events, thereby immersing them in narrated multimedia presentations about those events. To date, my students have produced a series of five situated documentaries based on past events at the Columbia University, Morningside Heights campus, on the Upper West Side of Manhattan, New York. The subjects of these situated documentaries include the 1968 student revolt or strike, the history of the campus as the mid-nineteenth century home of the Bloomingdale Asylum for the Insane, the half mile of tunnels honeycombing the campus, the history of early nuclear research at the campus (in the 1940s Enrico Fermi conducted research at Columbia, and even enlisted members of the football team to haul radioactive material through the tunnels), and the Edwin Armstrong story (Armstrong was a Columbia professor who invented FM radio, but ultimately and tragically committed suicide, at least partly due to a long-time struggle over the legal rights to his invention).

Conclusion

These technologies provide many important opportunities to improve news reporting during times of crisis. They can give reporters improved access to critical information and images at times of intense uncertainty. The public can get improved access to news and information, and perhaps one day will be able almost to relive past events placed in the context of where they once occurred. Together, these technologies can help provide vital information to citizens to help them make better-informed decisions.

But these inventions also raise troubling questions and concerns. Will news organizations use these technologies responsibly to improve the quality of journalism? Or, will they use them simply to improve audience ratings? Will privacy be crushed by smart news cameras that can not only watch but recognize us and track our every move (or might the media's use of such technologies counterbalance law-enforcement agencies' use or abuse of them—or merely add to the onslaught on privacy)? Will government agencies exercise restraint when closing the shutter on satellite-imaging systems during times of crisis?

No one knows how these questions will be answered. But only a continued public dialog and lively debate will ensure that the public interest is represented at the table when crucial and consequential decisions are made regarding the use of these technologies in news coverage.

5

Digital Photojournalism

Cheryl Diaz Meyer

SOMEWHERE AMONG THE TRILLIONS OF STARS in the southeastern sky was a satellite that was going to transmit my photographs from war-torn Afghanistan to the *Dallas Morning News* in Texas. That was my very basic understanding as I underwent intense training in the new digital technologies the week before my departure to cover the U.S.-led war on terrorism. It was three weeks after the terrible events of September 11, 2001, and the newspaper was determined to send its own people into the field and give its readers timely, in-depth news.

I quickly became adept in the use of my laptop with a NERA World Communicator satellite phone and in the use of a Sony digital video camcorder and two more satellite phones, which would transmit video shot by the writer, my willing multimedia partner Tracey Eaton. Eaton agreed to write, shoot video, and do live transmissions from the field to the nineteen television stations owned by our mother company, the BELO Corporation.

The satellite phone and the videophone system were all purchased specifically for our coverage in Afghanistan and were, therefore, the latest available technology.

My camera equipment consisted of two Nikon D1 H camera bodies, a 17–35-mm lens, a 60-mm lens, an 80–200-mm lens, a SB 28 DX Speedlight, eight camera batteries, two rechargers for the camera batteries, fifty

AA batteries, fifteen memory cards, a Macintosh G3 laptop computer, two laptop batteries, an AC power adaptor, a DC power adaptor, a power strip, and enough other electronic gizmos and widgets to numb the mind.

My boss, Director of Photography Ken Geiger, equipped me to face almost any technological challenge and still be productive. The worst-case scenario I could manage with my equipment involved a hardwire connection to a vehicle and power transmitted from a car battery.

As we launched our entry into Afghanistan from Dushanbe, the capital of Tajikistan, journalists were already exiting to replenish supplies. They told tales of generators destroying expensive equipment, inflated prices for services from drivers and translators, a lack of drinking water, and terrible dust.

We could deal with all of the above, but we thought to avoid the power issue by purchasing our own gasoline generator. I scoured Dushanbe for two days and paid a pretty price for a small new generator. Then we were off to face the unknown.

We were loaded down with ninety liters of water, one hundred pounds of videophone equipment, one hundred pounds of photo equipment and laptops, winter clothing, sleeping bags, food, cooking utensils, emergency kits, toilet paper, and our trusty generator.

The new technology seemed impressive until we had to carry it along with all of our supplies into the horrendous conditions of war and dust. On our second night in Afghanistan, we battled a dust storm in an old military tent as the winds tore through at 100 mph. We survived with the help of our videophone equipment cases, which held down the corners of our tent. Other journalists in the compound were less fortunate since their tents blew away with them asleep inside. No one was hurt. Plastic resealable bags saved our equipment from the dust that had seeped through cracks and zippers in our luggage. Dust, rough roads, and sticky fingers were the bane of our technological existence in Afghanistan.

Journalists found that Afghan soldiers and citizens were anxious for contact with the outside world and handheld satellite phones would go missing or codes would get jumbled as translators, guards, and others would try to contact relatives in Iran, Saudi Arabia, and the United States. Some journalists didn't think to bring their codes with them and would find their equipment nonfunctional.

The war in Afghanistan concretely redefined new boundaries for photojournalism. It was the first war in which all still photojournalists worked with digital cameras, transmitted via satellite phones, and were constrained only by the rigors of war. A few magazine photographers worked simultaneously with film and digital cameras since their deadlines came later.

The NERA satellite phone transmitted my photos of the dead and suffering, the refugees, the girl schools, and the bombing. Often, I transmitted a series of fifteen photos each day. Many other photojournalists struggled with smaller satellite phones, sometimes waiting three hours for a photo to move. Tensions were high, and, for some, the stressful job of covering a war was made nearly impossible by inappropriate equipment. Others arrived in Afghanistan with no satellite phones and paid fellow photojournalists $275 per image to transmit. Geiger said:

> There are two ways to think about satellite phones: You can either go light-weight and be tethered to slow data transmissions with great

FIGURE 5.1
A Northern Alliance soldier waves triumphantly to fellow soldiers as they advance on Konduz, the last Taliban stronghold in northern Afghanistan. (Photo by Cheryl Diaz Meyer.)

portability and voice communications, or you can suffer nine pounds of weight and carry around a laptop-sized satellite phone that will give you sixty-four-kilobit data rate, which is equivalent to half an ISDN line. That means a photo will move in about a minute. We go to produce multiple images with a more in-depth and expansive look at news or features.

A good satellite phone is not only necessary for transmitting photos, but also good for research and reporting. By accessing the Web with my satellite phone, I was using Inmarsat M4 high-speed data service, which allowed a direct connection to the Internet. I could see what my fellow photojournalists were doing and learn the latest news in Afghanistan and around the world. As a journalist, having a good satellite phone was a safety issue and a necessary tool for the job.

The videophone system that we carried to Afghanistan contained a Sony digital video camcorder, the 7E Communications Talking Head, two satellite phones with strong antennas, and a slew of cords and batteries that fit into two hard cases, fifty pounds each.

Eaton produced a video journal transmitted regularly via satellite phone and transmitted several live shots to television stations in Dallas, New Orleans, Seattle, and Portland, providing immediate news for our viewers.

"The videophone was valuable in terms of giving us a feel for the difficulty in Afghanistan before the fighting, and once it started, a powerful sense of what that fighting was like," said George Rodrigue, vice president of the BELO Capital Bureau.

Rodrigue emphasized, however, that regardless of the technology, content is critical. The chaos of the war when control was wrested in northern Afghanistan from the Taliban and passed to the Northern Alliance was our primary coverage.

Since neither Eaton nor I had worked with a videophone before our travels to Afghanistan, intense training involved engineers from BELO's Capital Bureau who were, in some ways, learning the equipment as they instructed us on its use. "When you send inexperienced people with new technology into the field, you need a lot of technical support back home," said Rodrigue. "It's very important to get skilled people working with them to polish the material sent home."

The videophone made by 7E Communications has been dubbed the Talking Head, and it fits into a case the size of three laptop computers.

FIGURE 5.2
Rebel soldiers witnessing Taliban soldiers surrendering in Afghanistan. (Photo by Cheryl Diaz Meyer.)

Two satellite phones and camera equipment fit into another hard case the same size. Eaton would take video footage of himself providing a news report on location or of a live incident in the field. In the evening, after writing his story and transmitting it, he would then set up the satellite phone antennas outside, dial into our offices in Washington, D.C., and transmit raw video. The camera would hook into the Talking Head, and the satellite phones would connect us to our home office via satellite. The Talking Head turned the video into data, and then a device on the other end would decode it back into video.

Rodrigue covered the Persian Gulf War in late 1990 and early 1991. He remembers typing his story on a typewriter or computer, giving it to the military to transmit, and having to wait for sometimes up to twelve hours for a transmission.

Later, he covered Bosnia during the fighting when no telephone networks functioned and stories were held for weeks. It was the satellite telex machines that finally allowed him and other journalists to transmit text stories via an Inmarsat terminal. The text was sent to a satellite, which connected to the Internet. "There were times when it was worth your life to be out on the street [looking for a phone]," said Rodrigue, a proponent of new technologies.

The Leafax negative scanner, introduced in the late 1980s, was the first machine that made it possible for newspaper photojournalists to transmit timely images from the field. It was the size of one medium hard case and weighed thirty pounds. The Leafax scanned negatives and created digital pictures using analog lines to transmit photos every forty minutes. The problem was that the quality of the transmissions depended on a good phone line, a scarcity in remote areas.

Soon, much smaller film scanners followed with laptop computers. Most recently, digital cameras have reached the quality and price that allow most newspapers to equip their photographers with the latest technology. At the *Dallas Morning News*, the staff of thirty photographers shoots with digital cameras; most have been issued laptop computers.

Nikon's newest digital camera, the D100, is a six-megapixel SLR aimed at both professionals and consumers. It is lightweight and offers more features than its professional counterparts, the D1H and D1X. It sells for about $2,000, less than a professional 35-mm film camera.

Canon's EOS D60 is the equivalent of Nikon's D100; it costs around $2,200. Both the EOS D60 and the D100 easily produce quality 16 × 20 prints.

The digitalization of photojournalism makes electricity and power even more critical to our work. Dirty gas in Afghanistan destroyed our brand-new generator within ten days, and the generator in turn burned several of our AC power cords. We could function temporarily with backup equipment, but not for long. Other journalists experienced complete computer meltdowns from generators and auto batteries that spiked to 350 volts and beyond. So, as the job of the journalist becomes more remote, it becomes necessary to not only learn the technology, but the equipment that will run the technology. A good fuel filter may have saved many journalists from trouble in Afghanistan, but that was still no guarantee given the dust.

Since the introduction of the Leafax film scanner, photojournalism has been careening forward to embrace the ever-increasing selection of tools that make the trade more transportable.

Fewer boundaries define photojournalism. The latest efforts of photojournalists involve still photography, audio, and video. Photojournalists are taking jobs as video photojournalists producing video footage for newspaper Web sites and television. Still photojournalists are taking voice recorders into the field so that multimedia presentations including photographs and sound are presented on newspaper Web sites.

Much debate surrounds these changes in the field of photojournalism. "Just because we can do it doesn't mean it's the right thing to do," said Geiger, who questions the increasing demands placed on photojournalists today.

"What do we prize most at the *Dallas Morning News*?" Geiger asked rhetorically. "Good photography. The level that we expect our photographers to perform precludes them from multitasking," he said.

Philosophical differences run deep and believers in either camp stand firm.

A handful of photojournalists around the country welcome the new technology, and some newspapers, including the *Dallas Morning News*, have committed staff photographer positions dedicated to pioneering efforts in the field of video photojournalism.

David Leeson, a veteran photojournalist of twenty-six years and a *Dallas Morning News* staff photographer, has been working full-time as a video photojournalist since the fall of 2000. Bringing a prize-winning career with him, Leeson calls himself a "DV journalist" or "digital videojournalist," a term he coined to describe the new breed of pioneers breaking off from mainstream still photojournalism.

In a field so new, even the vocabulary is still developing. Another term used to describe these technology adventurers is "platypus," coined by Dirck Halstead who teaches a workshop for still photojournalists venturing into video photojournalism. Platypus refers to the multifunctional role involving still photography, audio, video, and animation that Halstead believes is the future of photojournalism. The platypus metaphor is used because that animal appears to be a hybrid of a bird and a mammal, just as the multimedia journalist is a hybrid of two crafts: the combination of a still photojournalist and a video photojournalist.

"DV journalism is not yet defined because it is so young," said Leeson. "We know what it may become, but not necessarily what it is now."

He ventured to give a definition: A DV journalist uses the ethics, credibility, and high standards of still photojournalism to communicate in the language of sight, movement, and sound that same legacy upheld among newspaper photojournalists.

A growing number of video photojournalists propose that video journalism is the future of photojournalism and that newspaper still photojournalists will one day produce images for print and television.

"Tomorrow's photojournalists will not be using a 35-mm camera," said Leeson. "They'll be using a digital video camera. However, that camera in the future will be a hybrid camera because as resolutions continue to improve for digital video, so does the quality for the frame grab from that video." Presently, a video camera has enough resolution to produce a four-column image for the newspaper.

In his work, Leeson uses a Canon XL1 video camera, digital video editing software such as Final Cut Pro, and sound equipment including the Sony Mini Disc recorder and a variety of microphones.

As the field of digital video journalism expands and more photographers take on this hybrid endeavor, the demand for highly skilled photo editors will also increase, suggested Leeson. A photo editor who once may have edited thirty-six images from a roll of film may eventually edit thirty-six thousand images.

Multimedia journalism has found a home on newspaper Web sites. The *Washington Post* is known today for its successful Web site, washingtonpost.com. The site combines excellent photojournalism, stories, video, audio, and graphics in a rich mix of news and features that has become the paradigm of newspaper Web sites. Among photojournalists, it is known for its innovative approach to storytelling. Under intense deadlines, the staff produces multimedia galleries that incorporate audio such as interviews from subjects, music, and poetry.

In 1996, the *Washington Post* was preparing to launch its subscription-driven online news service, Digital Ink, when the Web burst onto the scene. The Web provided a forum that cost significantly less, would reach more people, and made production much easier.

In contrast to many newspaper Web sites around the country, washingtonpost.com is an individual business entity that employs almost two hundred people. The staff gathers news for the Web site and also uses material from the *Washington Post* newspaper.

"Washingtonpost.com pioneered the consistent use of quality photojournalism as part of the newspaper's Web presence," said Keith Jenkins, former director of photography for washingtonpost.com and presently photography editor for the *Washington Post Magazine*. Jenkins believed that the Web had equal potential for words and photos, but in its early stages, most newspapers considered the Web simply a way to catalog stories. Photos were viewed as bandwidth hogs.

Jenkins was driven to prove the viability of photos on the Web and was adept enough in the language to create pages that used photos extensively with text. One of the first few photo-driven projects was a story about a circus family written for the newspaper's features section. Jenkins repackaged the story into chapters that were introduced by thumbnail photos. More photos were then embedded into the body of the story. Although the story was very long, it received a record number of hits and overwhelmingly positive response from viewers and the *Washington Post* newspaper staff.

This led to experimental live coverage of President Bill Clinton's second inauguration in 1997. A three-person team consisting of a photographer, writer, and producer created a photojournal that was updated several times throughout the weekend's festivities. It was the first time that a newspaper had instantaneous news and photos on its Web site.

Today, washingtonpost.com's photo department is headed by Tom Kennedy, former director of photography at *National Geographic*. The staff of twelve to fifteen includes photographers, videographers, photo editors, and producers. "The *Washington Post* has made a substantial commitment to the Web that has led to an understanding that multimedia is an equal and essential component of online news," said Jenkins.

"I think newspapers will not change their nature," Jenkins continued. "They will essentially continue to be text products, but as technology improves, I think we will see more Flash-driven pages where text, photos, and audio are combined in more fluid and interchangeable ways."

At the University of North Carolina, Chapel Hill, Rich Beckman, professor and director of visual communication, focuses on teaching students the latest technology as well as the importance of journalism.

"We are empowering students to tell stories in an interesting and more diverse manner while still maintaining journalistic excellence," said Beckman. Along with photojournalism, students learn the value of audio in storytelling, learn to create presentations incorporating slide shows and audio of their work, and learn documentary multimedia storytelling as broadcast, photojournalism, multimedia, graphic design, and news editorial students combine efforts to produce high-end multimedia Web and CD ROM products.

In the future, Beckman believes that the job of the still photojournalist will still exist, but that a diversification will occur with multimedia journalists who will combine more than one skill. Multimedia journalists will understand the skills for audio, video, and still photography and know when it is appropriate to use each.

At The Poynter Institute in St. Petersburg, Florida, Kenny Irby heads the Visual Journalism Program as group leader, where among other activities he coordinates the Visual Edge Conference and Workshop. The events combine veterans and students of photojournalism who study two other forms of storytelling. They take video and audio equipment into the field and produce a Web site of experimental work.

According to Irby, photojournalism schools have the unprecedented challenge of meeting the core curriculum of photojournalism, writing, and ethics, in addition to teaching the multimedia skills required by the industry.

Irby does not justify or preach the need for new skills; he simply offers training to professionals and students who want or need to learn them.

"Because of technology, photographers can spend more time with subjects," said Irby. "You can deliver content more efficiently and quickly from remote areas, and that content can be multimedia-based. I'm not saying this is great—I'm saying, this is the reality."

The challenge of learning new technologies and poor training has caused degradation in the overall quality of photojournalism. Proponents of new technologies support experimentation believing that mastering them will eventually bring the quality of photojournalism to an even higher level.

"I think the work of photojournalists will be valued more because of the immediacy of the work, and the quality that photographers are able to produce will rise again," Irby said.

"In three to five years, I see photojournalism working in a much more dynamic way—in many ways, independent, more often in remote locations because of the digital technology advances and capabilities," said Irby.

Bibliography

Brian Cleary, "The Wire Service Way," at www.bcpix.com/wire.html.

Digital Photography Review, at www.dpreview.com.

History of AP photos, at www.ap.org/anniversary/nhistory/index.html.

Larry Larson, "New Technology Delivers Instant Reports," at www.poynter.org/content/content_view.asp?id=473.

Platypus Workshop reports, at www.digitaljournalist.org.

Washington Post, at www.washingtonpost.com.

6

Satellites, the Internet, and Journalism

Adam Clayton Powell III

Editor's Note: This chapter was adapted from a speech given to Canadian journalists gathered at the University of British Columbia.

From Science Fiction to Reality

NEW TECHNOLOGIES ARE HAVING A HUGE IMPACT on journalism and the way that journalists inform people about what is going on in the world. I want to show you something that you may not be aware of yet. When we initially started doing programs about this technology, we thought of it as science fiction; we thought, here is something that someday our grandchildren will be able to use. But it is now in daily use in large television and newspaper newsrooms and is a fascinating example of a technology that, like the Internet, originated in the military and spy agencies and is now in the hands of almost all of us news editors. I also want to show you how these new tools are affecting not only journalism, but also the whole debate around the role journalism plays in society today.

The Technology

We can now use pictures taken from satellites and combine this data with aerial photography, maps, and elevation data from NASA to create visuals for television news. This was all pretty expensive to do until about a year or two ago. Now you can do this on a laptop using satellite imagery that you can buy, allowing news operations to enhance their news coverage and show their audiences pictures from places they cannot get reporters to. Certainly there are many implications for privacy because people can now look down at anything—including into our backyards—from up in space. You can imagine how some U.S. agencies and other agencies around the world would like to control this technology, but since you can buy the images from Russian satellite companies, for instance, who are desperate for U.S. cash, the spy agencies are no longer in control of this technology.

Information Freedom Versus National Security

During the war in Afghanistan, the U.S. government bought up all the satellite images over that country at a cost of two million dollars a month. It had exclusive rights to all U.S. commercial satellite imagery and remote sensing of Afghanistan, so no news agency had access to it. A legal mechanism called shutter control was set up in the United States to deal with these types of issues. The U.S. government can say, "This is a matter of national security, and no one should be able to show these pictures," but the government did not use that law. In fact, we were all waiting for the government to invoke it; instead, it performed what some would call an end run: It simply bought up all the rights to the imagery so no one else could get it.

A Radio and Television News Directors Task Force in the United States had been negotiating with the Pentagon and other agencies for more access to the information, but since September 11, 2001, the news media have been getting less access to government information. The Federation of American Scientists is a nonjournalistic group that has pioneered the use of this technology, and they have put up on their Web site (fas.org) images related to arms control, treaty violations, troop movements—all the things that governments do not want to be made public. They also have a section on refugee camps, which has been one of the arguments made against the U.S. policy on Afghanistan.

Privacy

An important issue related to satellite images is privacy. As one editor in California said, "This [satellite technology] can look down into Jennifer Lopez's backyard, and while I'm not interested in that, the *National Enquirer* might be." "Inside Edition" or one of the tabloid TV shows might get shots of Jennifer Lopez sunning herself. In fact, using this technology the Freedom Forum demonstrated in Arlington that— if the light is right in space—you can actually look through people's windows. Intelligence agencies have been doing this for decades, but now we can all do it.

Buying Satellite Images

Foreign countries are using their satellites to sell images to American news organizations. The Russians want our currency, and so they are delighted to sell images from their spy satellites if you pay in non-Rubles. If you order a custom image, the press rate is anywhere from $300 to less than $1,000, but you have to be willing to wait a day or two. If you want the image right away, that can be really expensive, costing $25,000 to $50,000, because they have to use fuel to reorient the satellite. But the images available online and posted on the fas.org Web site are amazing: One page is devoted only to secret Russian nuclear test facilities.

Who has rights to this material? you might ask. Well, if you deal with say spaceimaging.com, they will e-mail you their terms of use. A major daily is granted up to two or three uses. For television there is a window of time for use. If you go to fas.org, you will see the source list for satellite images. Most of these sources are so anxious for attention or publicity that they will make very favorable deals with news agencies.

If you want a satellite image, a lot are available on spaceimaging.com. Space Imaging was founded in 1994 and makes use of the IKONOS satellite. A lot of images are given away, although the company does want credit in the cutline. If you need a custom image shot of a specific location, they can work with you on acquiring it. It is a fascinating tool that has been available to us for a relatively short time now.

There is another company, digitalglobe.com, that I want to mention. They have a satellite that went up recently, and it can officially image up to one meter. You can't see faces, but you can see people. A new satellite that spaceimaging.com and digitalglobe.com are putting up is supposed to go down another order of magnitude—down to a tenth of a meter. But they may not be allowed to do this because at a tenth of a meter you can begin to recognize people's faces.

Potential Problems

One editor in Sacramento said, "This is a great tool to catch the bad guys, but I don't want anybody looking at me." And that's the classic catch. There's another catch. The U.S. Navy spy plane that was forced down in China last year is a good example. One of the most famous sets of satellite images was of this plane because every news organization wanted a picture of that plane sitting on the ground. Whether it was directly from the satellite or from one of the wire services, everyone used the picture of the plane on the ground. One of the first pictures came from spaceimaging.com, and since this is not a precise science, somebody said, "Let's clean up the picture a little bit." Those of you who have been photo editors know what that means. So, they cleaned it up a bit, and when this image was compared to the image from the day before, it looked like a piece of the fuselage was gone. As a result, some news organizations assumed the Chinese were taking the plane apart. That's a pretty good story except that this appearance was caused by pixels they pushed too far trying to clean up the image, or pixelation error. As a result, however, people did stories about how the Chinese were taking the plane apart. It's an important point because how many of us are trained satellite image interpreters?

The Internet and Journalism

Now when we talk about globalization, we all say the Internet is global and will become more global; the media became more global—certainly the ownership is more global. We probably all use the Internet to gather information on breaking stories. I happened to be in Belgrade doing a BBC program when things were happening with Milosevic. As quickly as possible, I started just reading things from Web sites to see what people around Europe were doing. Classic BBC provided solid coverage, but then you start to see some other interesting things. I don't know if you know about the Institute for War and Peace Reporting (IWPR), an organization that facilitates on-the-job training for journalists in areas of conflict, but in England, it actually beat the *Guardian* online for a major journalism prize recently. But it's not a classic journalism source. If you write "according to IWPR," readers will ask, who? IWPR had a very good ongoing account that night based on first-hand reports from the Balkans. Mainly, they are trained for human-rights abuses and that sort of thing. So it's not classical journalism, but they are providing information that a lot of news agencies, particularly in the United Kingdom, have folded into stories. Established news sources have taken IWPR dispatches from iwpr.net.

We know that eventually it's all global. We happened to have somebody from Croatia at a conference we did in London. He asked, "Have you seen our campaign Web sites?" When we put those Web sites side by side, their candidates' Web sites were better than any of the U.S. candidates' Web sites, and completely done by college students, as opposed to high-paid consultants. We looked at one Croatian candidate's Web site. His campaign schedule was listed hour by hour—it must have driven the security people crazy, but you could click on the day or time to see where you might want to be. After the Croatian candidate gave a speech, within forty-five minutes his Web team would post excerpts in Real, Quicktime, and Windows Media. That's pretty amazing. So you have video of the candidate's most recent stop right there. By the way, his Web team also cached everything. The candidate's entire campaign was all there in video clips. I haven't seen too many candidates do that in other places around the world.

History Online

You can now also grab history online. There is a terrific site called archive.org. You can put in a date, and the Web site presents you with vast amounts of archived content from the Internet on that date. So if you're looking for something, or if you want a day-by-day account of what, for instance, happened in Florida, you can put in November of last year and just examine the Internet day by day in archive form. A lot of newspapers make a lot of money selling their archives. Who needs to pay for them, however, when one can go to archive.org?

There is also something called TVarchive.org, which presents twenty-four-hour-a-day coverage from twenty-odd television networks around the world, including Canada. So if you want to see what happened, you can actually put in September 11 or 12, 13 or 14, and it's all there. It will launch mpeg-4 video of what was going on. They have Chinese television, Russian television, Kuwaiti television, Al Jazeera, Iraqi television. There's a chilling sequence after September 11, 2001, on Iraqi television: They presented live shots of New York, interspersed with shots of the Rodney King beating, of race riots in the United States, of the bombing of Baghdad. Now I don't speak Arabic, so I don't know what the report was saying, but I can't imagine it was about what was going on in New York. Archive.org is a nonprofit organization that has devised a very, very cheap way of storing video using $300, low-end, bottom-of-the-line Hewitt Packard PCs, purchased one hundred at a time. If you go into their basement in San Francisco, you will see they have all the news broadcasts from all the major news outlets.

Immersive Media

The last example is about immersive media. It's not quite virtual reality, but very much like it. A lab in Los Angeles has a U.S. government contract from the National Science Foundation to be the engineering and research center for multimedia. The lab won that contract five years ago by saying that, at the end of the contract in 2007, they would have produced a device that would cost $500 and create for the user "an experience indistinguishable from reality." That's pretty good, right? No goggles, no earphones: It immerses you in an experience.

One of the applications that they're working on is an immersive car. If you look at your car radio using immersion technology, it becomes the ultimate stereo. It is also voice interactive: As you're listening to a report of President Bush's news conference, you can say, "Tell me more," and it will ask, "On what subject?" You say, "Well, I want to know more about Afghanistan," and it will start to play those pieces of the news conference. The program immerses you by placing your point of view; for example, it places you next to Helen Thomas so the president's voice would appear to be right there and you'd hear Helen Thomas's question next to you.

You can see some of the directions that this might take by 2007. Maybe people can visit a refugee camp or venture into the middle of some war zone. What happens when you do that? Some might think it would be great, a translating experience. But of course, editors will still decide what experience to transmit, what point of view to provide.

Another example of immersive technology comes from Fox Sports, which was interested in the technology for its potential application to Super Bowl coverage. The technology is still in development and was being tested on Super Bowl XXXIV in 2000. A man who has done a lot of work with movie and high-end sound design was consulted. He designed immersive three-dimensional coverage of the game, and those working on this project asked some interesting questions about where the fan should be. They began by putting the fan right next to the coach because then the fan could talk to the coach. Then they thought, No, let's give the fan the ultimate experience and put him or her on the field with the quarterback. You can do that with a lot of directional mikes, a lot of rendering, and a lot of extra mikes around the stadium. They could generate a live three-dimensional immersive sound experience of the game, and you would feel like you were on the field when the quarterback caught the snap. The quarterback was right there. They found when they did this in a room with three-dimensional audio and two big-screen televisions that all of this information had a pronounced psychological effect. The fans really do feel like they are on the field with the quarterback. The problem is that they're not used to being on the field next to a quarterback with a bunch of linebackers coming at them, and some people find the experience too intense. What is going to happen when producers of this technology start to use it to transmit from Afghanistan? What are the standards going to be? This is such a new area, but we have no idea what the conventions are going to be. To read more about this, see an article entitled "Super Bowl 3-D Audio Production Previews Challenges for 21st-Century Journalism" on the Freedom Forum's Web site (www.freedomforum.org/templates/document.asp?documentID=11370).

Now the issue is, how will you know if what you're seeing, or what you think you're experiencing, is real? Photographers, in particular, have taken the initiative concerning this. They're trying to create some kind of watermark or something that can't be altered on the original image; then, if the visual information were tampered with, the watermark would be broken. An unbroken watermark would ensure that you really are seeing or experiencing what you think you are. The Radio and Television News Directors Association has been talking about this with photojournalism organizations. I think most people think something has to be done because it has become too easy to alter images. Even with the best of intentions—like just cleaning up that satellite image of the plane sitting on the ground—you can create something that is very different from reality with this immersive technology.

Our most profound problem is that, based on everything we know, technological innovation is speeding up. We are entering a period where the rate of change in the tools we're working with as journalists is greater than ever before, and authenticity, attribution, and verification are more important than ever because the tools are so powerful.

Bibliography

For more information about some of the issues discussed in this chapter, visit the following Web sites. Powell cautions that a number of URLs may have changed.

For articles about high-resolution commercial satellite images:

Arvidson, Cheryl. "Journalists Denounce Government 'Shutter Control' Idea For Satellite Images." *The Freedom Forum Online,* at www.freedomforum.org/templates/document.asp?documentID=11944.

Kees, Beverly. "CBS News Technologist Urges Government to Share More Satellite Data." *The Freedom Forum Online,* at www.freedomforum.org/templates/document.asp?documentID=12026.

Powell, Adam Clayton, III. "High-Resolution Commercial Satellite Images Introduced." *World Center,* October 13, 1999, at www.freedomforum.org/templates/document.asp?documentID=11380.

For articles about the use of the Internet in nontraditional settings:

Powell, Adam Clayton, III. "West African Journalists Explore E-Mail, Text Tools for Online Reporting." *World Center,* June 10, 1999, at www.freedomforum.org/templates/document.asp?documentID=5844.

Powell, Adam Clayton, III. "Internet Use up Sharply in Peru." *FreedomForum Online*, March 13, 2000, at www.freedomforum.org/templates/document. asp?documentID=11876.

Other Articles

Broad, William J. "Now It's Torn Up, Now It Isn't. Photos Differ on Spy Plane." *New York Times*, April 11, 2001, A12, at query.nytimes.com/gst/ abstract.html?res=FA0E15FB3D550C728DDDAD0894D9404482.

Powell, Adam Clayton, III. "All-News 'Radio' Goes High Tech, Interactive with Voice Command, 360-Degreesound, Web Search." *World Center*, November 23, 1998, at www.freedomforum.org/templates/document.asp?documentID=11172.

Powell, Adam Clayton, III. "Countering Counterfeit Press Releases—And More." *freedomforum.org*, January 25, 1999, at www.freedomforum.org/ templates/document.asp?documentID=11360.

Powell, Adam Clayton, III. "Super Bowl 3-D Audio Production Previews Challenges for 21st-Century Journalism." *World Center*, January 29, 1999, at www.freedomforum.org/templates/document.asp?documentID=11370.

Ghana Web Sites

Ghana Broadcasting Corporation (government) (www.gbc.com.gh)
Ghana News Agency (government) (www.gna.com.gh)
Independent radio stations:
Joy 997 FM (www.joy997fm.com.gh)
Gold 905 FM (www.gold905fm.com)
Choice 1023 FM (www.choice1023fm.com)
Vibe FM (www.vibefm.com.gh)
Newspapers:
The *Graphic* (www.graphic.com.gh)
The *Chronicle* (www.ghanaian-chronicle.com)
Lusaka Post (independent) (www.zamnet.zm/zamnet/post/post.html)

Other Web Sites

CBS News technology site (http://cbsnews.cbs.com/network/htdocs/digitaldan or www.gizmorama.com)
CBS News Disaster Links resource page (www.cbsnews.com/network/htdocs/digitaldan/disaster/disasters.htm)

Federation of American Scientists (www.fas.org)
(Note: FAS's Public Eye satellite program, formerly at www.fas.org/eye, has
  been dropped from the FAS Web site. John Pike, who organized it, left FAS
  and started a new center, Global Security. You can see the resources there on
  the Global Security Public Eye Project Web site.):
Global Security Public Eye Project (www.globalsecurity.org/eye/index.html)
Space Imaging (www.spaceimaging.com)

7

Social Movements and the Net: Activist Journalism Goes Digital

Melissa A. Wall

THE END OF THE TWENTIETH CENTURY saw the rise of a new era of activism—from college students working against sweatshops to communities of faith joining human rights groups to oppose Third-World debt to activists around the globe challenging the corporate control of the world's economic, environmental, and political agenda. With this upsurge in advocacy has come the use of the Internet to help sustain these movements. Every organization, it seems, no matter how small, has a Web site and an e-mail list. In some cases, it is difficult to separate the movement from the medium as some groups seem to exist only online, often as larger and more sophisticated entities than they are in reality. Decentralized, nonhierarchical, chaotic, anarchic—these terms are used to describe the activist networks and the Internet itself (Arquilla and Ronfeldt, 1998; Castells, 1996.)

In tandem with these changes has come an increasing concentration of mainstream media. A handful of companies such as AOL Time Warner and Vivendi now own much of the world's media. The intersection of these changes—a rise in activism, the increasing pervasiveness of the Internet within the activist toolbox, and the rise of giant media conglomerates that appear to favor the status quo—has prompted an increased emphasis on the production of news by social-movement activists.

It seems clear that activist journalism has greatly benefited from the Internet, which has provided advocates a new means of creating and distributing their own versions of events, while combining that information with mobilizing messages intended to prompt immediate responses. But what exactly distinguishes digital activist journalism from mainstream practices? Are there any models out there now that can help explain these differences? What do alternative practices tell us about the future of digital journalism?

Social Movements

The African American struggle for civil rights; antiwar demonstrators' fight against the Vietnam War; the women's movement; the global struggle against South Africa's racist policies of apartheid, which segregated and brutalized the black majority: All of these are examples of social movements—groups of people seeking to create social change. A social movement embodies a challenge to the powers that be, seeking to right what its members see as societal wrongs or injustices. While many are progressive, not all social movements favor positive social change; the racist white militias that gained power in the 1990s across the United States are also considered a social movement.

While a social movement encompasses a range of individuals and groups, among their key players are organizational actors called social-movement organizations. Such organizations often orchestrate or fund movement activities providing mobilizing structures, background research, and philosophical grounding. For example, during the Civil Rights Movement, black churches and other organizational entities were heavily involved in organization and mobilization. The Farm Workers' Movement to gain fair pay and better working conditions for agricultural workers was spearheaded by a social movement that eventually became the United Farm Workers' Union.

While most of these groups have certain behaviors in common, their news-creation practices are of interest here. In the past, some experts have argued that social movements couldn't exist without mainstream media coverage and that seeking this coverage often influenced how movements behaved. But if today's activists can "be the media," as they put it, then this dependence may be waning and movements' own powers as journalists waxing.

Activist Journalism

From feminist newspaper editor Susan B. Anthony to DIVA-TV AIDS video activists, movement actors have often also served as journalists for their movements (Kessler, 1984; Streitmatter, 2001). Activists believe that mainstream coverage tends either to ignore their issues entirely or to cover them in a biased and unfair manner, belittling and discrediting their movements. Activist journalism seeks to provide the movements' side of the story, which is often strikingly different from what the mainstream media present. They seek to mobilize constituents, prompt action, and create movement identities. Historically, their task has been complicated by the fact that activist media have been underfunded and have struggled with issues of financing without ceding their mission to advertisers or other corporate interests. Production has sometimes been infrequent and distribution slow and difficult because of the expense involved. With the rise of the Internet, activists believe they can solve some of these fundamental problems such as the expense of long-range distribution and even of content production (Downing et al., 2001; Atton, 2002).

Online Providers of Activist News

Just as mainstream media now create their own online versions, so have activist publications such as the *Nation* and the *Progressive* posted Web editions of their publications. However, digital journalism has made its greatest impact in the introduction of new models for providing activist news. The two most common models today are (1) organizations that are *activist news outfits* focusing on progressive movement-oriented news and (2) organizations that are *movement organizations* focused on specific issues (e.g., the environment, women's issues) and that also provide news.

Online activist news differs from the mainstream in several significant ways. First, their news tends more toward commentary or opinion. These groups are generally unconcerned with reporting the other side, in part because they believe that the dominant voices in society, such as those of corporations, are already favorably represented in the mainstream media and are often not speaking for the good of average people. They see themselves as providing an alternative point of view, one that is often missing in the corporate-controlled media. Second, they tend toward the reproduction and redistribution of information that has already been reported and published elsewhere. While this is not always true, in many cases they are operating more like the *Utne Reader* or a progressive version of the *Reader's Digest*, both of which tend to scour other publications for their material. The origins of these articles vary, but their main sources are alternative media (often print publications), nonprofit advocacy or social-movement organizations, and progressive voices found in the mainstream media (most often on the op-ed pages). Third, these sites tend to be text oriented, placing less emphasis on photographs, illustrations, video, or audio. Like their dependence on reproducing material, this tendency seems to stem primarily from a lack of resources. Fourth, activist news sites generally are nonprofit and do not rely on advertising for funding, but are financed through donations, sponsorship, or foundation money. Finally, activist news often strives to offer a broader range of voices, aiming to open up discourse on controversial issues to underrepresented groups.

Activist News Outfits

Activist news outfits are those groups whose primary mission is to deliver news of an alternative nature. An assessment of these groups reveals three main types:

1. Professional news services, such as Alternet or Common Dreams, that either operate solo or in partnership with a nonprofit sponsor
2. Sponsored news services, such as WorkingforChange, that partner with corporate entities
3. Grassroots news services run by volunteers and open to all contributors, such as the IMCs or Infoshop

Professional News Services

One of the best examples of an activist professional news service is Alternet, an alternative online news magazine sponsored by the nonprofit Independent Media Institute. Alternet began operating in 1998, offering news, opinion, and investigative pieces from perspectives it believes are not offered by the mainstream news system. Articles come from alternative newspapers and magazines, mainstream publications, and nonprofits. In collecting news items and reports, the San Francisco–based outfit partners with more than two hundred media and advocacy organizations.

Alternet calls itself an "infomediary" because it scours the alternative press every week for its content. (It also accepts a handful of freelance articles.) This control over its content—one cannot simply post a story on Alternet—lends it a professional quality. The gatekeeping—the process of selecting some stories and leaving out others—is performed by a staff that draws heavily from the alternative print media. Executive Editor Don Hazen, for example, is a former publisher of *Mother Jones*. The result is a thoughtful compilation of perspectives. Among its content foci are "Enviro/Health," "Rights and Liberties," "DrugReporter," "MediaCulture," and "Global Affairs."

Alternet is financed by the sale of articles through a syndication system with more than 150 member publications and through grants and contributions from groups ranging from the Ford Foundation to George Soros's Open Society. The only advertising on its site is a Working Assets button. In addition to providing news, it also provides places for visitor feedback, including letters to the editor and online forums for discussion and debate among site visitors about the issues presented on Alternet. The forums pose various questions to which visitors respond, such as "Since the War on Some Terrorism isn't doing too well at actually finding terrorists, let's just kill Iraq instead; at least it can't run away. What's wrong with this picture?"

Sponsored News Services

An example of a sponsored news service is WorkingforChange, sponsored by the progressive telephone company Working Assets. It too runs articles from activist and mainstream media, as well as articles from social-movement organizations (SMOs). Its editor, Christopher Scheer, was formerly a foreign/national editor at the *San Francisco Examiner*. While progressive voices are heavily promoted (it includes a Web radio show hosted by feminist media activist Laura Flanders and regular columns by leftist commentator Alexander Cockburn), the site also relies on a more commercial model than most others of this type.

The site features advertisements for Working Assets, Ben and Jerry's ice cream, and links to a sister site called ShopforChange. The column "What We Love" lists favorite items compiled by staffers that are available for purchase by site visitors (e.g., one week a columnist wrote about her favorite independently owned magazines with links to a page that allowed you to subscribe immediately to the magazines; another week a staffer recommended music for toddlers). In addition to opportunities to shop, the site also encourages political action with links to pages that explain issues and include prewritten e-mails that can be sent from the site to elected representatives.

Grassroots News Services

Grassroots news services differ from the others in that one needn't be a professional journalist in order to get one's story posted. Instead, anyone who has something to say can cover a story or contribute a comment. Among the most innovative groups in this model has been the IMC movement, which was created in response to the World Trade Organization ministerial protests in Seattle in November 1999. Media activists believed that the corporate media would not tell the whole story of the protests, so they set up a Web site to give the other side with a particular emphasis on volunteer participation and on representing the marginalized voices of women, people of color, the economically disadvantaged, and so forth. Around a hundred other IMCs have subsequently been launched in the United States and around the world based on this model.

On IMC sites, anyone can participate in the production of news by posting a story, photograph, video, or audio—be it originally reported or pulled from another source. Each posted item can then be commented on by anyone, which sometimes results in a series of comments from different people whose discussion in this public forum becomes part of the story. The IMC further differs from the other models in that it is more anarchistic, usually more raw and immediate, and sometimes less reliable. The site also differs in that it tends to carry more than just text, including video, audio, and photographs. More so than the other services, the IMC is intimately connected with social movements, particularly the Global Justice Movement that came to life in Seattle. Most IMC sites prominently list upcoming protests, demonstrations, and other activist events.

Social-Movement Organizations

In addition to the creation of online activist news sites, a second model is evolving. Here, certain SMOs or advocacy groups such as Greenpeace are also using the Internet to become specialized news distribution centers or, in some cases, participants in alternative news networks. (For a case study of this phenomenon see Wall's "Battle in Seattle" [2002]). For the most part, the news being produced or distributed by these SMOs is particular to whatever the organization's specialty is and generally more in-depth than that of the activist news sites because it concentrates on explaining a single or handful of issues. Thus, information that would be raw material distributed to mainstream media (e.g., research reports or position papers) is offered unfiltered as information for direct access by site visitors. In addition, these groups have set up Web pages that work as in-house news centers, distributing the SMOs' news releases as well as articles from other sources about the topic. Items range from stories taken from mainstream or alternative media to reports, releases, or other information being shared by other SMOs. Two different models are as follows:

1. Network sites collecting news and information for and about SMOs such as OneWorld or WTOWatch
2. Specific SMO's own Web sites that distribute news on specialized topics related to that particular SMO's mission such as Global Exchange on global justice issues or Greenpeace on environmental issues

Network Sites

One example of a site that serves as a networking host for SMOs and their issues is U.K.-based OneWorld.net. Launched in 1995, it is devoted to news and information about human rights and sustainable-development issues. It seeks to distribute information representing marginalized people who are the objects of human-development projects or otherwise impacted by development entities. To do so, OneWorld has become what it calls a "public-interest portal," partnering with more than 1,250 groups around the world that work in human-development areas such as the environment and debt relief. Content comes from a mix of reports and articles from SMOs and activist-oriented media. For example, a page devoted to the Middle East conflict between Israelis and Palestinians includes reports from a range of organizations: media groups such as Inter Press Service to Arabic Media Internet Network, as well as SMOs such as Amnesty International and Grassroots International. OneWorld has eleven centers around the world from Amsterdam to New Delhi to Washington. Each center helps organize the site's pages and is owned by a nonprofit in that country (with the exception of the Global South centers). Thus, OneWorld United Kingdom is owned by the Panos Institute, while OneWorld Italy (Unimondo) is owned by the Fondazione Fontana. The parent organization provides funding. In 2001, OneWorld also signed an agreement with Yahoo!News to provide content on social justice issues.

OneWorld's interactivity opportunities depend on which section you visit. The site include links to what it calls "ethical shopping," a site for feedback about the site itself, and opportunities regarding some issues to sign up for e-mail digests or to submit resources (e.g., reports, position papers, links to other sites) on whatever the focus issue is. At the newer OneWorld TV site visitors can post video clips that "introduce new evidence, personal testimonies or a different perspective." Clips range from the story of a Soweto mother dealing with AIDS to Amnesty International's collected video testimony from West Bank Palestinians.

Issue SMO Sites

San Francisco–based Global Exchange is a nonprofit research, education, and action center focusing on global justice issues and is noted for its campaigns, such as convincing Starbucks to carry fair-trade coffee, which meets certain environmental standards, but also pays peasant farmers a living wage. Its Web site operates as a highly specialized news outlet. For example, on its campaign pages devoted to the conflict in Colombia, it includes "news updates," articles about the conflict collected from mainstream and alternative media as well as SMOs and advocacy networks. A link to a *New York Times* article might follow a link to a press release from the Traditional Authorities of the U'wa Association. In this way, the SMO is creating new hierarchies for information in which mainstream information is displayed as equal to that coming from SMOs, alternative media, and other advocacy sources. Other pages provide Global Exchange's own analysis. Global Exchange also encourages visitors to act, providing pages to help send faxes, write letters, or telephone government representatives, as well as opportunities to sign up for issue-oriented e-mail lists or to volunteer to work with global justice issues.

Summits and Crises

Activist news being created and distributed by those groups seems to have the most impact when the world's attention is focused on the issues they are concerned with. This tends to happen with large crises; consider the responses to September 11, 2001, in the United States or the increased violence in the Middle East that rose to new heights in 2002. Other key moments that focus the world's attention include global events like United Nations summits or other gatherings of world leaders such as the World Trade Organization meetings. These events often provide a physical place to gather and a specific, dramatic issue around which to organize. Increasingly, citizens are using these summits to call for a voice in the processes, a voice that they argue their governments provide only to giant corporate players, not to the ordinary people who are most impacted by these decisions.

The coverage of these summits by the mainstream press is often considered flawed and inadequate. As SMOs have come to play an increasing role in the events, their ability to create and distribute information has been part of their rising power. They help frame debates, mobilize constituencies, and speak for the grassroots. The borderless nature of these events fits with the borderless nature of many globally oriented social movements around the world today and may explain why these events are the highpoints of reporting and information distribution for activist media. Indeed, indymedia is believed by many to be at its best at these summits when it marshals its resources for a brief, focused spurt of global reporting. Many indymedias have been created in connection with key events in various locations around the world ranging from the U.S. Democratic Party Convention in Los Angeles in 2000 to the increased violence in Palestine in 2001 and 2002 after the September 11, 2001, bombings.

Conclusion

Activist digital journalism holds both promise and peril. The perils center especially on issues of access. While most of the organizations described above are committed to enhancing the disempowered, such voices—especially globally—do not yet have equal access to the communications technology to participate fully. Additionally, there is the issue of who will ultimately control everyone's access to the Internet—even within rich countries such as the United States. Increasingly, a handful of companies seeks to own the means of entering and navigating the Web and other digital communication forms. In the end, these business interests may delegitimize digital activist journalism by making it extremely difficult to find or by simply ignoring it.

The promise lies in the ability of activist news outfits and the social movements they support to create an alternative information sphere that provides news, reports, and mobilizing information. Because such an information sphere is anchored by different values, it challenges us at the most basic level to reconsider what is "news." In this virtual space, news is no longer a daily report of events produced by trained, "objective" professionals within a long established range of practices and routines. In this new digital activist sphere, objectivity in the news is seen as impossible to achieve; instead, reporters are often movement members who share movement values—producing commentary and think pieces more often than not about their issues of concern. In addition, this sphere anticipates an active audience that will not only make sense of complicated events, but act upon its opinions about such news. In this sphere, news is not a commodity whose value rests on how much one can charge for it; instead, news is seen as a public good to be used freely by citizens and redistributed, not hoarded. Ultimately, digital activist journalism represents a new stage in social-movement communications and in the definition of news itself.

Bibliography

Arquilla, John, and David Ronfeldt. "Preparing for Information-Age Conflict; Part 1: Conceptual and Organizational Dimensions." *Information, Communication and Society* 1(1) 1998: 1–22.

Atton, Chris. *Alternative Media*. London: Sage, 2002.

Castells, Manuel. *The Rise of the Network Society*. Vol. 1, *The Information Age: Economy, Society and Culture*. Cambridge, Mass.: Blackwell Publishers, 1996.

Downing, John D. H., et al. *Radical Media: Rebellious Communication and Social Movements*. Thousand Oaks, Calif.: Sage, 2001.

Kessler, Lauren. *Dissident Press: Alternative Journalism in American History*. Thousand Oaks, Calif.: Sage, 1984.

Streitmatter, Rodger. *Voices of Revolution: The Dissident Press in America*. New York: Columbia University Press, 2001.

Wall, Melissa A. "The Battle in Seattle: How Nongovernmental Organizations Used Websites in Their Challenge to the WTO." In *Media and Conflict: Framing Issues, Making Policy and Shaping Opinions*, edited by Eytan Gilboa. Ardsley, N.Y.: Transnational Publishers, 2002.

8

Digital Government and an Informatics of Governing: Remediating the Relationship between Citizens and their Government

Paul W. Taylor

A Potential Transformation

THE NEW MILLENNIUM OPENED AMID a blush of bipartisan optimism in state capitols across the United States. At the crest of the longest sustained peace-time economic expansion, with the Internet fueling both innovation and speculation, state treasuries were flush with revenue surpluses, general unemployment rates were in the low single digits, and programs intended to help public-assistance recipients move into the workforce appeared to be working, at least for the job-ready.

Overlooked in the hyperbole of the dot-com-fueled new economy were the combined impacts of three significant shifts in the previous decade on the act of the governing. The first was the withdrawal of all but the most stalwart news organizations (or, in some cases, determined reporters) from state capitols. Many media organizations (often conflicted internally about their own digital-journalism strategies) saw little downside to closing their bureaus in state capitols— official sources were regularly eschewed in favor of real people who lived the story; audience research bolstered the view that political coverage was not interesting or important in the day-to-day lives of the now sovereign reader or viewer; and the consultant-driven, formulaic government-waste stories were just as easily done by parachuting into town just before sweeps periods.

The second shift was an institutional response by government itself to the Year 2000 (Y2K) computer date-field problem, which represented both significant risk and opportunity for the public sector. The risk was the legal liability that came with the potential loss of vital public services or public accountability. Governments came to Y2K remediation without public confidence on their side, owing to earlier failures of high-profile information technology (IT) projects in states across the country and despite success rates for government IT projects that closely approximated those in the private sector, which has the dual advantage of failing outside of public view and not using scarce public funds.[1] With their credibility on the line, governments needed to be successful with Y2K remediation (the largest software maintenance project in history). Moreover, there needed to be a return on investment for the extraordinary allocation of taxpayer funds that extended the serviceable life of the technology systems underlying government services.

The third shift was a change in public behavior. Self-help books were firmly ensconced on best-seller lists, home-improvement warehouses proliferated in aid of a new generation of do-it-yourself home owners, and discount online brokerages catered to a new class of individual investors who had fired their financial advisors in favor of doing it themselves. They were ready for, even demanding, self-service government.

This chapter is focused on the outworking of these three factors around the rise of the commodity Internet—initially in the popular imagination, and ultimately as a mass medium in its own right. It begins with a conceptual treatment of the architecture and function of the state capitol dome as a useful analogy for thinking about government's presence on the Internet (which disintermediates conventional media from the citizen-government relationship while expanding convenience and choice). The discussion continues with a fifty-state overview of the progress made during the first wave of digital government from 1997 to 2002. The chapter concludes with a broad outline of what comes next by introducing a framework of governing informatics.

The call for an informatics of governing is rooted in the recognition that (1) those who founded governments had a more comprehensive view of the enterprise and its purpose than do many of those who seek to transform it through digital government, and (2) the state portal may be the only seat of government that most citizens will ever know.[2] The latter observation came only with maturity born of experimentation and hard-earned experience. The early initiatives owned as much to dot-com excitement as to a philosophical commitment to governing through technology.

The New Seat of Government

Architecture can be awe inspiring. Beyond obvious examples, which range from the National Cathedral to Hoover Dam, consider the state capitol. It is too often taken for granted by those who work in and around it. But see the domed capitol through the eyes of first-time visitors—tourists, grade school students, and the like—and one is reminded of its visual impact.

As the official home of the state flag, the state seal, and a portrait gallery of leaders past and present, the capitol building is high on symbolism. It is also, by design, high on function. It is the place where the people's business gets done—supported by a network of operating agencies that stand behind the capitol building with a reach extending across the state. The combination is at once compelling and comforting—just watch the first timers approach the grand edifices and enter these civic temples.

In the sometimes-overused parlance of the Internet, the capitol is the original public-sector portal. As such, it is a useful standard bearer for those who are building twenty-first-century government.

One Government

The state capitol represents a declaration of intent that the people in a geographically defined space, which spans multiple cities and counties, will act together as a single entity, sharing the burdens and the benefits of community. At best, such a community is bound together by both practical considerations of cost reduction and mutual aid, and by a big idea that is sometimes captured in the state motto—Alaska's "North to the Future," Kansas's "*Ad astra per aspera*" ("To the stars through difficulties"), and New Hampshire's "Live Free or Die" come to mind.

The big idea for the state Internet portal is to provide and support the kind of government that was imagined by the people who first chiseled those words into stone at their respective state houses, without the constraints of time or space that characterized the earlier era. The Internet collapses geographical barriers, making government available at the time and place of the citizen's choosing.

The branding of government Web sites continues to be a matter of considerable debate, as if the issue were really about the people and organizations within government. In fact, the state capitol has long had outposts that do the work of the state under the symbols of the state—the flag and seal being the most prominent among them. The historical precedent contends for a common look and feel on the Web because it provides important cues to people using the services. Just as the government buildings that stand behind the capitol carry common symbols and operate similarly, there is a reasonable expectation that government Internet applications standing behind the portal should look, feel, and function in like manner.

The communities of interest that are forming around the design and deployment of the citizen's online experience are stretching a common pixel layer across the many previously discrete departments of government in favor of a functional view. Sector-specific "one-stop" or "no-wrong-door" initiatives have been amplified through options for personalization and customization at the portal level in some states and localization or regionalization in others. Whatever the path, it is vital to maintain anonymous paths to online information and through as many online services as possible. Such an approach can be useful in maintaining an appropriate balance between public access and privacy and ensuring that citizens are able to move as freely about their portal as they are to move around their capitol.

A Fair Hearing

People come to the state capitol to petition their government. They have an expectation of timely response and a hope of favorable resolution. There are expectations of the petitioner, too—places where he or she cannot go and conduct that is restricted out of concern for the well-being of others.

Here too, the portal is becoming a surrogate for those who would have otherwise made a trip to the capitol. Their petitions must be directed or redirected to the appropriate parties, receipt of inquiry acknowledged, and a response sent. The practices developed ad hoc and are now being codified to ensure consistency and thoroughness in handling citizen contact. As for the expectations of the petitioner, there is a growing body of case law that suggests government is well served in defining through policy the intended use of its Web presence up front or face the daunting challenge of managing a public forum after the fact.

Visitors with routine questions about the capitol rely on docents for answers or directions as to where to go for more information. The docent's function online is handled at the portal by a combination of frequently asked questions and a search function. Docents and search engines should both be chosen with care—the good ones are worth much fine gold, the bad ones are more trouble than they are worth.

Trust

State houses are the subject of a recurring and important debate between the competing legitimate interests of openness and public access on one hand and providing appropriate security on the other. It mirrors a similar debate about the Internet itself. The fundamentally open architecture of the Internet stands in opposition to attempts to limit, censor, and even contain data traffic. Yet, it is recognized as proper and useful to hold some data securely, to close access to some applications, and to harden some data flow.

It is worth remembering that many of the people who built state capitols had taken a page from early bankers halls, which were built to engender trust. Banks of that era featured impressive columns outside, marble floors and counters inside, and an invincible-looking vault as their centerpiece. Form met function—the buildings held most thieves at bay, while instilling confidence in customers that their deposits would be safe.

Issues of ego notwithstanding, the builders of state capitols used many of the same architectural cues to instill the sense that important things happened there and that these institutions were worthy of doing the public's business.

As portals mature to include advanced applications that deal in personally identifiable information, they carry all the expectations of history and the challenges of doing business securely in the sometimes-dangerous place called the Internet. Portals must confront the maxim of Internet security: mutual distrust until proven friendly. The onus is on the operators of Web properties to prove themselves friendly. To that end, in an online-privacy statement written in plain English, people should be informed at the point of collection how their information will be collected and used, if at all. The promises made by the privacy statement can only be kept through vigilant attention to security, ensuring the integrity of the underlying infrastructure and applications, and aligning implementation and operation with the related security policies and practices.

Differentiated from Private Sector

Who does one trust online? Early on, many in the public sector assumed that the dot-com top-level domain was the most credible suffix on the Internet. That's changing. Webmergers.com, a research and real estate company for Internet properties, estimates that at least 862 dot-coms have failed since January 2000.[3] By contrast, every government in the country is still standing.

Governments in the United States are doubly blessed in the naming protocols of the Internet with two top-level domains: dot-gov and the less lyrical state.us. The capitol is called the capitol because that's what it is. As a rule of thumb, government Web properties should be called what they are—dot-gov. These dedicated governmental Internet domains have one other advantage over their commercial sibling—maintenance. The government domains are essentially a monopoly, registered in perpetuity to the jurisdiction they name. In contrast, expiring dot-com names, including those registered to public entities, are often intercepted by the adult-entertainment industry. The address people have come to trust, as an extension of their government, can suddenly become home to unwelcome graphic surprises.

The People's Space

There is still important work to do in expanding functionality, hardening security, and becoming more disciplined in content management on most public-service portals. But the thrill of launching a portal, and the experience of building it, are now just stories people like to tell.

The men and women who built the capitol buildings did not stay around. Their work was done. They were builders, and they went on to the next project. A different group of people, with different skills and even dispositions, came in behind to maintain them. Engineers worry about the cracks in the foundation, while crews shine the brass railings, polish the oak desks, and tend to the flower gardens out front.

It is important to recognize a similar, healthy transition where portals are concerned. Just as the capitol is the most carefully maintained real estate in a state, the portal needs that same level of care and attention. The operational staff of the portal, like craftspeople who maintain the capitol, have more direct influence on the citizen experience of the government than the elected official.

The Big Closed Door

The heavy, ornate steel doors that lock shut each evening help keep the capitol building secure while nobody is around. But they also telegraph the message that government chooses the time and circumstances for doing business with the public.

Portals create the opportunity to improve on that part of government's reputation and responsiveness. The portals and the applications that stand behind them are available around the clock. Government should no more have people clicking through unattended Web properties in the middle of the night than have them wander through the capitol after the staff have all gone home.

Enter the 24/7 customer-care function—a live e-mail and phone center where technical and business questions about applications are triaged and resolved. Through an iterative process, the lessons learned from live interactions with users are rolled into the frequently asked questions, thereby continuously strengthening the self-service option for the next wave of visitors. The tracking of customer-care inquiries indicates a disproportionate volume of business being done with the state at the very end of the day—11 P.M. and later. It is then that people have questions and really appreciate answers. Consider the feedback of one woman who resolved an issue in the middle of the night through Washington State's customer-care center: "Wow, I never knew government could act like this."[4] Indeed, acting like that is the point of digital government.

A Noble Enterprise

The state portal may be the only seat of government that many citizens ever know. Public trust in government has to be earned over and over again, and it is won or lost with each interaction between citizens and their government. Even the selfless responses by many public servants to the September 11, 2001, terrorist attacks on America were only able to sustain above-average public-approval ratings for less than a year.[5]

It is sobering to walk the halls of state capitols. The architecture, the statuary, the inscriptions all reflect the aspirations of the people who dared to carve their values and dreams into stone. The permanence, the elegance, and the grandeur of these public spaces points to a faulty design assumption in much of what has been built in the government Internet space to date—it reflects a reductionist (even trivialized) view of government and has not been built to endure.

Good Government and the Innovator's Dilemma

By 2002, the optimism of the new millennium had been pummeled by terrorist attacks on U.S. soil, the implosion of the once Midas-like technology sector, political and corporate scandals that shook public confidence, and a severe economic downturn that emptied state coffers, creating revenue shortfalls of a magnitude never before seen by career fiscal forecasters in state government. Together, these changed circumstances tested the commitment of political subdivisions to becoming digital. While important progress was made during the first generation of digital government, the inertia was considerable.

Government in the United States is built on more than two hundred years of history and tradition. The processes that grew up around that rich history first confronted automation some four decades ago. All of this makes government particularly prone to the Clayton Christianson's "innovator's dilemma," the formulation that good organizations tend to gravitate toward enabling technology, while often missing the impacts of disruptive technologies.[6] The former allows for incremental improvements of existing processes, while the latter—as the name suggests—disrupts tired, old processes in favor of a change in the order of things.

Christianson assesses no blame for the blindsiding of the disruptive, noting that good managers earn that reputation by dedicating themselves to keeping organizations running smoothly—not breaking them. Public-sector managers see the world rather differently than their private-sector counterparts, due in no small part to the negative incentives that characterize public service—the reward is for the avoidance of errors, not always for the attainment of good. Yet, Christianson warns of external disruptive forces that abhor the status quo the way nature abhors a vacuum. His conclusion has been popularized in a single sound bite— cannibalize yourself before someone else does.[7]

Cannibalizing Government Services

The Internet and market forces encroached on what had long been considered the exclusive purview of government—public records, information, and even routine transactions between citizens or businesses and public entities.[8]

In an era in which the so-called killer app was the holy grail of the commodity Internet, the stuff of government appeared to be a leading candidate. The Pew Research Internet and American Life Project reported that fully half of Internet users looked for government information online, slightly fewer than the percentage that checked the weather, but more than those who made a purchase, booked travel, or visited an auction site.[9]

Similarly, a twenty-eight-country survey by the Australian firm Taylor Nelson Sofres reported that online government is second only to buying books as the highest-use category among Internet users worldwide.[10]

From Monopolist to Competitor

The marching orders given by the then-governor of North Carolina were typical, directing agencies under his control to "dot-com government." The theme echoed across the country, with the early adopters taking widely different approaches to the new common objective. Some states pursued electronic, or e-, government internally, insisting that the Internet was the new core competence of government. Other states relied on traditional IT integration and technology companies to move them to the Web. Still others relied on a new generation of third parties—Internet startups that focused on high-volume transactions with intrinsic economic value.

Such transaction also attracted other players, including media companies such as AOL Time Warner, MSN, and Yahoo! that had portal strategies of their own and positioned themselves as an indirect channel for the conduct of routine government transactions.[11]

The rise of the indirect channel juggernaut effectively redefined state portals as nonexclusive access points to the applications that are behind them. As media properties, the channel providers have unrivalled reach. They draw eyeballs from places that census takers never thought to look. For their part, governments are the authoritative source of identity and licensure—and provide an elaborate suite of services to everybody, including those well outside the private sector's target demographics.

The strategic importance of driving up page views at the state portal pales in comparison to the value of driving large volumes of routine traffic to Internet applications, where the cost per unit of service is a fraction of that for conventional service delivery. The greater the traffic, the more government, channel providers, and citizens alike benefit.

It could well be that there is no need to reconcile friends in the state-portal-versus-indirect-channel debate. Channel providers and government each have something the other wants, which should allow them to come to the table as relative equals.

Even with this combination of resources and competitive pressures from other states and the private sector, digital government remains a matter of unfinished business. Research by a number of third parties provides a view of the landscape at the end of the campaign's first five years.

A study by Indiana University, Bloomington, on behalf of the Price-WaterhouseCoopers (PWC) Endowment for the Business of Government identified three waves in the evolution of state Web portals. The initial wave was a group of "dressed-up search engines" and an index of governmental links. The second wave "increased user control," moving beyond brochureware with "advanced search capabilities" and "enriched content." The third wave allows data interrogation and panagency access to online governmental applications, among other features.[12]

Using data from the 2002 Digital State Survey,[13] the Center for Digital Government populated the categories identified in the PWC Endowment study. The overlay indicated that 100 percent of states had satisfied the first-wave criteria and fully 90 percent had developed their virtual front door into a second-wave portal. It is noteworthy that, for all the years of effort and discussion that attended the introduction of online government, the portal is a very new entrant into public-service delivery. Specifically, most second- and third-wave state portals had been launched (or relaunched) in the previous two years.

Forty states (80%) were also providing access to online applications and data interrogation, the primary characteristics of the third wave. However, in considering the advanced features included in the third wave—customization, dynamic push, calendars, and instant messaging—and elements of what may compose a future fourth wave (secure gateway and 24/7 customer care), the build out in the states fell to below 50 percent.

Assessing the maturity of state Web properties should not be limited to technical features alone. The Digital State Survey results indicate that, in the majority of cases, state portals are accompanied by Web-specific policies and practices intended to demonstrate responsible stewardship of information and services provided online (see table 8.1).

TABLE 8.1
Characteristics of Mature Portal Policies and Practices by States

Maturing Portals	Implementation (%)
Privacy statement	94
Security statement	76
Universal design/accessibility policy	82
Intended-use notice	40
Out-of-band contact/address information	80
Inquiry response protocol or policy	52
Content update procedures and dates	48
Transaction receipts	58

Source: Center for Digital Government, 2002.

A key attribute of third-wave portals is access to online applications through which citizens and businesses can conduct transactions with public entities. The suite of services available online varies widely from state to state. As table 8.2 shows, even among the ten most widespread applications available through state portals, only four have been deployed in more than three-quarters of the states. The table excludes partial implementations—those that require human intervention or another channel to complete a transaction.

There is useful experimentation with so-called democracy portals—and even online voting pilots[14]—that address the individual as a citizen, not just a customer or taxpayer. That said, the universe of common applications is dominated by routine transactions that are useful, but do not conjure up patriotic themes.

Still, in keeping with government's relationship with the individual, e-mail links to elected and appointed officials, as well as to civil servants responsible for specific programs, are ubiquitous at state portals and almost three-quarters (71%) of city and county Web sites.[15] However, government can do more toward respecting its online users without any additional investment or technology. The practice of providing privacy and security statements, which is the dominant trend among state portals, is less mature at the application layer (see table 8.3). Roughly half of the top ten applications across the fifty states have such notices, despite the fact that personally identifiable information is much more likely to be collected and displayed at the application layer than at the portal itself.

TABLE 8.2
The Most Widespread Online Applications Offered by State Governments

Common Applications	Full Implementation (%)
Online job search	98
Unclaimed property search	96
Legislation tracking	94
College admissions	94
Court decision lookup	70
Sex offender lookup	68
Vital records	68
Professional licensing lookup	62
Business tax filing	58
Business licensing lookup	56

Source: Center for Digital Government, 2002.

TABLE 8.3
Privacy and Security Notices at State Portals
and the Most Common Online Applications

Presence of Privacy and Security Statements	State Portal (%)	10 Common Applications (%)
Privacy statement	94	55
Security statement	76	46

Source: Center for Digital Government, 2002.

Likewise, an examination of most widespread application types in the nation's one hundred largest cities and one hundred largest counties indicates a widely uneven deployment of online services by local government (see table 8.4).

A separate survey by the International City/County Management Association (ICMA) of four thousand of its member jurisdictions indicates that requests for services such as pothole repair (31%), requests for public records (29%), the delivery of those records to requesters (19%), registration for the use of recreational facilities (13%), and the submission of permit applications (9%) are the most widespread online offerings available through the Web sites of political subdivisions of all sizes across the country. The ICMA survey also indicates that three-quarters (74%) of its respondents have a "Web page"[16] or, to use the PWC categorization, a first-wave portal.

A view of digital government informed by the number of deployed applications suggests that much has been done—and that there is much more to do. To date, two-thirds deployment is the high watermark for Web-delivered services at the state level and one-third at the local level. The still-latent potential to transform the cost of government funda-

TABLE 8.4
The Most Widespread Online Applications by the
100 Largest City and 100 Largest County Governments

Common Applications—County	(%)	Common Applications—City	(%)
Property tax lookup and pay	31	Online job search	39
Online job search	26	Parking ticket payment	31
Bid registration	19	Online bidding (RFP)	23
Vital records	15	Water bill payment	14
Building permit applications	9	Sanitation service/schedule	8

Source: Center for Digital Government, 2002.

mentally requires deployment of common, effective applications, the design of which requires careful attention to the underlying and invisible value chain. Only through reengineering the value chain—not just processes—can government mine the costs out of the existing order of things and meaningfully address the structural financial dilemmas it faces.

The Path Forward

Absent any other direction from the political leadership, years' worth of work remains to complete the build out of the current generation of online services. The demands of governing, of course, will not allow public agencies to catch up before introducing surprises and new challenges. Such challenges require of government a quality that might best be called institutional improvisation, a capability that includes, but is not limited to, what we have conventionally understood as digital government. Improvisation is, by definition, at odds with the deliberative nature of the legislative and executive branches of government. The inherent friction here presents a daunting challenge with which public institutions must grapple internally, or risk the blunt force caused by the external forces of cannibalization, such as citizen initiatives and plebiscites. Ironically, the United States, which has invested both financial and political capital in electronic government, may be ill prepared to deal with the success of the campaign—the ultimate impact of which may well be a collision of democratic objectives and republican constitutions.

The next step is not simply to finish the build out of online applications or the infrastructure that supports them. Both are necessary but not sufficient for governing effectively in the twenty-first century. The analogy to the capitol dome is helpful in thinking about the path forward.

Capitol campuses are home to institutions born of the constitution, which is the original covenant that bound together a geographically defined community of communities. It is uniquely American that state capitols are typically not found in the largest population centers, meaning that most of the country is governed from the periphery.

Given that structure, governments have built a network of brick and mortar outposts in communities across the state, while preserving control at the headquarters, as is frequently required by state constitutions. As we contemplate the future of governance, the historic role of governing from the periphery may more naturally and more effectively be performed through the network.

Seen in that light, digital government is the means to a larger end—digital governance. Figure 8.1 illustrates the wider landscape—and the impact of digital government and digital governance on the core functions and capacity of government.

The first cluster is around digital government, the original drivers of which were expanded citizen convenience and choice, while allowing the government to drive costs out of existing processes. The digital government sphere includes the service offerings discussed in this chapter—routine transactions in which the individual is seen primarily as a customer and the service is seen primarily as a commodity. Digitally delivered units of service have advantages for providers and consumers alike—routine information and service delivered consistently, reliably, quickly, and cost-effectively.

If digital government reaches from government to communities, digital governance is rooted in the life of communities—reinvigorating the public square through technology. A number of observers attributed the change in public priorities to the attacks of September 11, 2001, a catalyst for an anticipated reversal of the purposeful isolation that had characterized the decades just past.[17] Changes in geopolitics, technology, and the economy have been instrumental in connecting people back to local communities, shifting the focus from Wall Street to Main Street and the town hall.[18] Ironically, 40 percent of states have, on the advice of counsel, developed intended-use policies for their portals, which specifically define them as being for official purposes, not a virtual public square.[19] To meet the expectations of digital governance, these portals may have to become exactly that which these states attempted to prevent—a public forum.

There is anecdotal evidence of citizens using the instruments of digital government in attempts at digital governance. There are examples of people using the pothole-repair Web site to provide input into a discussion of transportation policy or budget priorities. Faced with this apparent mismatch, jurisdictions are finding ways to triage such input—routing it from operational staff to the policy or executive office. It is also recognized that such input should not be viewed as a

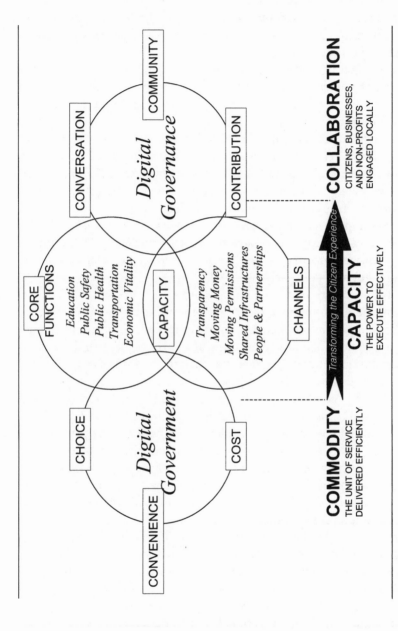

FIGURE 8.1
Toward a model for governing infomatics.

mismatch and that government ignores such input at its own peril. Rather than viewing sophisticated customer relationship management (CRM) systems as a means to track pothole repairs, perhaps CRM is of greater value in managing government's conversation with citizens about the future of their respective communities. In an optimistic scenario, a timely and responsive interaction with government on issues of interest and concern helps to keep citizens engaged and contributing to the life of the community.

The shift from commodity to community, coupled with structural difficulties in state budgets, forces a careful reexamination of the core functions of government. It is useful to differentiate the finite number of core functions from the hundreds of processes that have grown up around them. These processes are at the heart of the innovator's dilemma, where the exclusive focus is on incremental improvements of existing processes through enabling technologies. In contrast, a functional focus allows for the possibility of transforming or even eliminating dated processes through the introduction of disruptive technologies.

Finally, digital government and digital governance increase both government capacity and the demands against that capacity. Government's capacity has conventionally been defined in terms of human and physical resources, supported by a heterogeneous technical infrastructure that has been developed over the last forty years.

The Internet extends the existing technical infrastructure, while adding new channels for conducting the people's business in new ways. Governments have statutory responsibilities to conduct themselves in the "sunshine," a colloquial reference to public-disclosure and public-meeting laws. The portal, related Web properties, and the state or local public-affairs cable-television channels enable government to be transparent—that is, to meet the letter and spirit of disclosure laws—in ways that were not imagined when the requirement was codified.

Moreover, many of the core functions of government rely on moving money and permissions in substantial quantities. Paper processes are ubiquitous in government because, in large measure, the printing press has had a five-hundred-year head start in reshaping the world. Moving money and permissions is more naturally, more effectively, and more cost-effectively accomplished in a bit-based environment than an atom-based world. The government's challenge, and its opportunity, is to correct the accident of history that Johann Gutenberg was born five centuries before Tim Berners-Lee.

A Model of Governing Informatics

Words matter. Journalists understand that better than most. They also understand the difficulty of common expressions that are inadequate in capturing the breadth, depth, or subtlety of a big idea. The problem is further complicated by over and imprecise use, causing even useful terms to fall out of favor. Such are the nontrivial problems facing what has alternatively been known as e-government or digital government in the battle for political and public mind share.

Clearly, the challenges of transforming government functions through technology are weightier than the infrastructures (conceptual, political, and technical) created to support initial forays into electronic government will support. Everything "e-"—the once ubiquitous prefix for commerce, business, learning, and government in the much-ballyhooed new economy—lacked the heft to deliver on its promise. The term *digital government* allowed for a more comprehensive view of governing through technology, but in common usage it has not been sufficiently differentiated from the term *e-government* to maintain a clear distinction. If the act of governing is the central issue, then perhaps digital governance would focus our attention on the larger, more significant task. The intended meaning of digital governance is clear enough in context. Without it, however, it can become confused with the mundane internal task of overseeing of an organization's IT policies and projects.

It is instructive to see how naming conventions have developed at the intersection of other disciplines and IT. In the hard sciences and medicine, a critical mass has formed around the term *informatics*—that is, human-machine communication, or the science and art of turning data into information. Witness, for example, the breakthrough work in bioinformatics, where DNA and IT meet. Similarly, health-care informatics is the discipline that has grown out of the intersection of patient care and IT.

What of the social, or so-called soft, sciences? Specifically, can political science and the study of public administration generate a comprehensive informatics of governing? It can and it must. The ability to do the public's business and earn the public trust requires a new understanding of the interface between citizens and the act of governing.

A recent CIA report suggests that in the coming decades, governing will only get harder: "Governments will have less and less control over flows of information, technology, diseases, migrants, arms, and financial transactions, whether licit or illicit, across their borders. The very concept of 'belonging' to a particular state will probably erode."[20]

In that context, a model of governing informatics would be future oriented and challenge driven. A theory and practice of governing informatics necessarily begins with a renewed understanding of the constitutionally defined social contract between the governed and those who govern. It is the place where digital government becomes digital governance and provides a comprehensive view of governing through technology that is well suited to the spirit of the age.

In accepting the 2002 Randolph W. Giuliani Leadership Award, New York Finance Commissioner Arthur Roth used language that has, regrettably, fallen out of common usage in this country. He accepted the award on behalf of those who "ministered to the physical and financial needs" of the people of New York before and after the skyline changed. Commissioner Roth believes the post–September 11, 2001, environment has reminded us that, at its core, "government needs to serve."[21]

Three other "posts" underscore that same need—postboom, postnomadic, and postmodern—the long-term impacts of which on governing, while profound, have received comparatively little comment.

Postboom is the generation that has grown up in households where PCs were just another appliance. For them, the Internet is their first choice for commerce, conducting research, and connecting with a sense of community. Significantly, the first wave of this demographic cohort became eligible to vote at the turn of the century.

Joel Kotkin and Susanne Trimbath, writing in the *Los Angeles Times*, identify the current era as postnomadic, noting a return home for those who had been distracted from the priorities of "family, faith, and community."[22] It follows that they are much less likely to cede decision making about these rediscovered priorities to unseen civil servants or even elected officials operating outside of public view. Moreover, postnomadics see network connectivity as a utility. They have embraced the Internet as the power behind the newest generation of labor saving and other gadgetry in good times—and, more significantly, turn to it to check on the safety of loved ones in times of emergency. They expect the Internet (and the online government services that ride on it) to have the availability and reliability of electricity and tap water.

The third constituency is the loosely aligned postmoderns, the most vocal of whom have taken to the streets of Seattle, Quebec City, and Washington, D.C., among other cities, to protest globalization and the excesses of capitalism. Many have rejected the institutions of modernity and are, at best, suspicious of government. Importantly, however, they view the Internet as the last best chance for democratic renewal of America's republican institutions.

Postboomers, -nomadics, and -moderns will undoubtedly bump into each other in the electronic public square while contending for what may ultimately be irreconcilable worldviews. The hopeful and compelling sign is that they are all nesting in a networked world. In the main, they share a sophisticated view of the Internet—using it for communicating, commerce, creating community, and more than a little agitating. Government fails to at least match that level of sophistication at its peril. The other caution is to these constituencies themselves, to recognize that such democratic empowerment may conflict with the republican design of the constitution unless all parties can agree on those things that remain self-evident.

The CIA's stark prediction about government's loss of control in a society of free agents changes the rules for governing. We need a model of governing informatics to transcend the technology, people, processes, and functions of government to focus on the highest-stakes civil prize of them all—an idea. Namely, the idea of a participatory, representative democracy.

Acknowledgments

The author thanks Dennis McKenna, Cathilea Robinett, Steve Kolodney, and Todd Sander for their input on the ideas reflected in this chapter, and Joe McGavick for his thoughtful review. Mark Struckman, Janet Grenslitt, and Lauren Atlas contributed to the data analysis, collection, and tabulation reported here. Portions of this chapter have appeared in print as columns and essays in *Government Technology* magazine.

Notes

1. Capers Jones, *Software Assessments, Benchmarks, and Best Practices* (Boston: Addison-Wesley, 2000).

2. The term *Web portal* has been overused without precision. Definitions tend to focus on technical functionality. Simply put, a portal is an opening, and *Software Development (SD) Times* quotes a software executive who defines a Web portal "as an opening to knowledge." Edward J. Correia, "Are Portals the Workspace of the Future?" *Software Development (SD) Times*, May 1, 2001, at www.sdtimes.com/news/029/story6a.htm.

3. Webmergers.com, *Internet Shutdowns Report*, September 2002, at www.webmergers.com/data/article.php?id=64.

4. Washington State Information Services Board, *Digital Government Plan: Transforming the Relationship between Citizens and Their Government, Release 2* (Olympia, Wash.: Washington State Department of Information Services, 2001).

5. National Public Radio, the Henry J. Kaiser Family Foundation, and Harvard University's Kennedy School of Government, *Tracking Poll*, August 2002, at www.npr.org/news/specials/civillibertiespoll2/index.html.

6. Clayton M. Christiansen, *The Innovator's Dilemma: When New Technologies Cause Great Firms to Fail* (Boston: Harvard University Press, 1997).

7. Jerry Useem, "Internet Defense Strategy: Cannibalize Yourself," *Fortune*, September 6, 1999: 121–134, at www.fortune.com/fortune/information/permissions/pdf/internet_defense_strategy.pdf.

8. David Danner and Paul Taylor, "Principle and Practicality: Funding Electronic Access to Washington State Government Information," *Journal of Government Information*, 24(5) 1997: 347–359.

9. Pew Internet and American Life, The Rise of the E-Citizen, April 3, 2002, at www.pewinternet.org/reports/toc.asp?Report=57; Getting Serious Online, March 3, 2002, at www.pewinternet.org/reports/toc.asp?Report=55.

10. *Government Online Study*, Taylor Nelson Sofres, Sydney, Australia, 2001, at www.tnsofres.com/gostudy.

11. William D. Eggers, "The Invisible State," *Government Technology*, February 2002, at www.govtech.net/magazine/story.phtml?id=8097&issue=02:2002.

12. Craig L. Johnson, Diana Burley Gant, and Jon P. Gant, "State Web Portals: Delivering and Financing E-service," PriceWaterhouseCoopers Endowment for the Business of Government (New York City: PriceWaterhouseCoopers, January 2002), at endowment.pwcglobal.com/pdfs/JohnsonReport.pdf.

13. The Digital State Survey is a fifty-state assessment of government use of Internet technologies in the service of citizens, originated in 1977 by the Progress and Freedom Foundation (PFF) and conducted jointly by PFF and the Center for Digital Government since 1999.

14. In addition to experimentation with online voting in selected locales, data from the Center for Digital Government indicates that fully 90 percent of states publish voter pamphlets on the Internet, over three-quarters make searchable election-filing databases available online, and two-thirds post campaign finance information.

The center has also identified two significant clusters of activity in changing the nature of the polling place. First, optical scanning of full-face ballots has emerged as a mainstream activity in place at polling stations in 80 percent of states. Two small states use optical scanners exclusively for tabulating ballots, while another eleven states use the technology to handle three-quarters or more of balloting. Second, the PC may have finally earned legitimacy in the polling place. In January 2002, an all-software, PC-based election system became the first of its kind to meet Federal Election Commission qualification standards to perform public elections in the United States. Importantly, touch screen, ATM-like, and Web-based technologies have established a significant presence in twenty-one states. The new digital devices support large-print, multilingual, and audio transcriptions of ballots, which bring the promise of more voters casting ballots secretly and with integrity.

15. International City/County Management Association (ICMA), *Electronic Government Survey* (Washington, D.C.: ICMA, 2002); Center for Digital Government (CDG), *Digital State 2002* (Folsom, Calif.: CDG, 2002).

16. ICMA, *Survey*, 2002.

17. Robert D. Putman, *Bowling Alone: The Collapse and Revival of American Community* (New York: Simon and Schuster, 2000).

18. Joel Kotkin, *The New Geography: How the Digital Revolution Is Reshaping the American Landscape* (New York: Random House, 2000).

19. Center for Digital Government, *Survey*, 2002.

20. U.S. Central Intelligence Agency (CIA), *Global Trends 2015: A Dialogue about the Future with Nongovernment Experts*, NIC 2000–02, Washington, D.C., December 2000, at www.cia.gov/cia/publications/globaltrends2015.

21. Arthur Roth, Prepared remarks on the receipt of the Rudolph W. Giuliani Leadership Award at the 2002 New York State Executive Leadership Institute, Fort Orange Club, Albany, N.Y., September 23, 2002.

22. Joel Kotkin and Susanne Trimbath, "Behold the Post-Nomadic Economy," *Los Angeles Times*, August 4, 2002, at www.newgeography.com /LAT_Post_Nomadic_Economy.htm.

9

Online Medical Communication among Peers: The Net and Alternatives to Traditional Journalism

Patricia Radin

Editor's Note: Patricia Radin explores a form of information sharing that supplements traditional sources of news and information with electronically mediated peer communication. Health information is one of the most popular genres of information that Internet users seek. There is a wide range of online health information to be found, much of it posted by traditional news media organizations. But people who share interests in a disease because they have it (or know someone who does) can also be sources of information and support to each other, using the Internet as their medium of communication and community building.

I'M 21," THE WEB MESSAGE SAID, "and I just got back from the scariest doctor's visit of my life."

The posting, from Seattle, appeared in a Nova Scotia–based online breast cancer discussion group, bca.ns.ca, late one evening. The writer had just learned she had breast lumps and was terrified at the prospect of cancer.

Within twenty minutes, answers to her posting began to appear: advice to get it checked out, reassurance that the disease rarely appears in one so young, a description of the biopsy process, links to online medical articles, and a kindly pledge that "even if it is breast cancer, there are a lot of us survivors out here, and we will help you through this."

This collage of support and information, assembled by strangers around the world, is an example of a new type of niche communication—online, mutual-help, medical communication among peers, a blend of personal anecdotes, advice, encouragement, and mostly amateur research. Because of its global reach and many-to-many configuration, the Internet can provide spaces for people with common concerns to find and communicate with each other in this new way. As a result, there are thriving online discussions among people with serious chronic conditions such as diabetes, multiple sclerosis, leukemia, organ transplants, and depression. Participants give advice and support; sometimes they locate and comment on research papers, news articles, and other information sources relating to their shared concerns. Essentially, active participants of these online communities all become content providers, and at the same time they are passionately loyal niche audiences.

In the case to be examined here, a communication space was opened up in December 1996 when a volunteer for Breast Cancer Action Nova Scotia (BCANS; bca.ns.ca) launched a Web site that included a forum. Although it was aimed at a local constituency, it was of course globally available, so Americans, Europeans, Australians, and others soon joined the Canadians. The forum began a growth spurt in 1998 that reached about one thousand visits a day by 2001. According to the site's server log, the visits have consistently lasted an average of thirteen minutes—comparable to the time people spend with their daily newspaper or TV news (exclusive of commercials).

Why do increasing numbers of people turn to the Internet for medical information? How credible is the information? What challenges or opportunities does online health communication pose for traditional journalism? This chapter, looking especially at the BCANS Web site, will address those questions from a case study perspective.

In this chapter, the words *health* and *medical* are used alternately to refer to the different dimensions of a medical patient's informational needs. "Medical" information may include specific research findings, therapies, products, statistics, or procedures relating to a particular condition. "Health" is a broader term that refers to a condition of physical and emotional well-being—coping with everyday discomforts, finding emotional support. The phrase *information seeking* itself is inadequate to describe the reaching out that occurs as people seek not only answers to their questions, but also empathy and caring tailored to their individual situations.

This chapter will review the literature on the communication needs of people with breast cancer; it will assess online information seeking; it will look in particular at how the Nova Scotia forum constructs medical information; finally, it will consider the implications of peer medical communication for the field of traditional journalism.

Communication Needs of Breast Cancer Patients

The young woman's message about her "scary" doctor's visit illustrates the shock, confusion, and urgent need for communication that typically assail someone who faces serious illness. An eloquent description of the newly diagnosed patient's feelings appears in the booklet *Breast Cancer Online*, which was collectively written and edited by members of the Nova Scotia–based online community (Reeve and Wagner 1999, 5):

> Strange and ominous words like CT scan, bone marrow biopsy, mastectomy, reconstruction, chemotherapy, radiation, and prognosis are suddenly thrust in her direction as if she is supposed to embrace them with ease and move forward gracefully. Emotional, physical, spiritual, and mental capacity are stretched to the limits of tolerance, then repeatedly asked to give some more. Nothing feels ordinary, safe, or clear. (Orina Mann)

Breast cancer is a particularly dreaded disease because of the widespread incidence (at least one in eight women in the United States will get it); aggressive, risky treatments; unpredictable outcomes; high mortality rate; and potentially slow, painful final stages. Despite its unique characteristics, the communication-seeking pattern I will discuss, although focused on breast cancer, is likely familiar to others who have faced other sorts of bad medical news.

Breast cancer information seeking plays two roles: It aids decision making at a time when a patient must choose among unpleasant options that will have uncertain results, and it reduces anxiety by allowing patients to do something for themselves. On the practical side, information seeking is "a key moderator between perceived threats of the disease and the likelihood of taking action" (Johnson 1997, 9). Dr. Jerri Nielsen, a physician with an ominously fast-growing breast tumor, who was stranded at the South Pole in 1999 with no possibility of rescue for eight months, wrote a poignant explanation of the emotional aspects of information seeking. She was the only doctor at the American base. She drilled her physician in Illinois with e-mailed questions. She performed self-biopsies as best she could and sent pictures of the tissue samples to Illinois via videocamera, trying to learn the precise parameters of her condition. She persisted in trying to gather information even after selected chemotherapy drugs were air-dropped to her in a daring off-season run, ending most of the decision making that was possible in her case. She explained, "I needed to hash out the issues, pick the scenario down to its bones, and then *reinvent it in a way that I could live with it*" (Nielsen 2001, 230, italics added).

While Nielsen, with her medical training, sought solace in facts, the typical patient craves emotional support as well. Personal accounts report wave upon wave of emotional crises as a suspicion is confirmed, decisions are faced, treatments are endured, family life is dramatically altered, and a long, deep shadow falls across the future (e.g., Batt 1994; Colomeda 1996; Gee 1992; Kushner 1975; Lorde 1980; Middlebrook 1996; Moch 1995; and Wilson-Hashiguchi 1995). Receiving good emotional support often has measurable medical benefits (Roter and Hall 1997, 186; Spiegel et al. 1989), as well as "better adjustment, better coping, higher self-esteem, and improved acceptance of the illness" (Kurtz 1997, 11). Good communication has been associated with greater longevity and lower perceptions of pain in terminally ill breast cancer patients (Spiegel et al. 1989; Spiegel 1997), as well as less depression and anxiety and better coping skills in people with cancer (Spiegel 1994).

This thank you message, posted from Pennsylvania to the BCANS forum, illustrates how helpful appropriate responses can be:

> I posted earlier in the week that I would be starting treatment and was getting nervous. I just wanted to thank you all for your sound advice and loving reassurances. Anytime I started to feel anxious I would go back and read your posts and it would calm my fears.

The typical experience that patients have with medical teams seems far less satisfactory: "Survey after survey has shown that many hospital patients feel that they have been inadequately informed" (Ray and Baum 1985, 44). Lerman et al. found that 84 percent of ninety-four American breast cancer patients who were surveyed reported "difficulties communicating with the medical team" (1993). For their part, surgeons acknowledge this shortcoming in communication skills, according to an Australian survey (Girgis et al. 1997).

Like so much of medical communication, cancer communication is very specific to the individual's medical situation and emotional needs at a given moment, and these can vary greatly. To illustrate the complexity of a diagnosis, the American Cancer Society Web site lists ten different types of breast cancer, five stages of the disease, four levels of cell aggressiveness, and 260 relevant medical terms, from ablative therapy to xeroradiography. Factors that may affect an individual case include not only the aggressiveness of the cancer, location of the tumor, and whether the disease has spread but also the patient's age, treatment decisions, personality, physical condition, economic situation, peer group, family and job situations, the capabilities of her health-care providers, and sheer luck. Against this huge matrix of medical possibilities is overlaid the individual's particular pattern of emotional and informational needs, generally corresponding to the following stages (Luker et al. 1995, 1996; DeGrasse, Hugo, and Plotnikoff 1997; Degner et al. 1997; Deane and Degner 1998):

1. *Concern:* A person becomes concerned about developing a disease, possibly after reading an ad, learning of a close relative's diagnosis, seeing a TV program, or noticing a suspicious symptom. This is a preliminary, low-key information-gathering stage.
2. *Prediagnosis:* Alerted to a possible disease, a patient undergoes tests and waits for the results. Support during this period is scanty (De Grasse et al. 1997), although anxiety is intense (Degner et al. 1997).
3. *Diagnosis:* When disease has been confirmed, a patient's first thoughts are typically, Will I survive? For how long? Under what circumstances? (Butow et al. 1996). During this period, where there is still sketchy information, she may need to arrange an absence from work and deal with family care-giving issues. Along with these stresses comes dread of the risks and miseries of surgery (Bilodeau and Degner 1996; Street and Voigt 1997).

4. *Postsurgery:* Now that the tumor, surrounding tissue, and lymph nodes have been analyzed, the medical team can piece together these clues and try to predict what comes next. Unfortunately, when this momentous information is delivered, the survivor is likely to have difficulty in absorbing the implications (Ray and Baum 1985). One woman described her horror like this (Middlebrook 1996, 2): "[M]y surgeon's words, 'invasive carcinoma,' bounced like overripe tomatoes off the hospital walls and splattered us blood-red with fear."

5. *Recovery and follow-up treatment:* This period brings great emotional and physical challenges as the survivor undergoes harsh radiation or chemotherapy. The resulting misery is, sadly, quite a different scenario from the steady improvement that patients expect after most hospital stays. Everyday aches or pains may trigger alarm in a breast cancer survivor who is "always waiting for the other shoe to drop" (Cooper 1988, 44). In-person support groups are offered at this stage, but only a small percentage of patients attends them.

6. *End-of-life issues:* Discovery of a metastasis is often the beginning of serious contemplation of death. It is a sign that previous treatment has not brought about a cure and the cancer has spread. While the disease may still be manageable for many years to come, the patient now may feel her time is bounded. Considerations at this stage include decisions about how much further treatment to endure and how to organize her affairs. Along with this may come an urgent search for alternative treatments or clinical trials. The Internet, essentially a searchable database of billions of files, excels at this type of brokering, bringing together a specialized treatment and a small set of potential patients.

Overall, Johnson (1997, 7) finds that the questions raised by people with cancer cover a broad range: Besides diagnostics and treatment, they seek details about scientific discovery, alternative approaches, continuing care, rehabilitation, and even political decision making that affects cancer research and care. They turn to various sources, but Johnson found (in a pre-Web era) that "women preferred doctors most, friends/family and organizations about the same, and media least for general information." He notes that this is ironic, since (in pre-Web studies published from 1974 to 1995) they in fact reported obtaining most of their information from their least preferred source, the mass media (77–78).

With the generally unsatisfactory state of cancer communication, the stage was set for the emergent Internet to fill a persistent need.

Online Health Information

In 1996, *Health Online* by physician and cancer survivor Tom Ferguson urged readers to buy computers, sign up with an Internet service provider, and start exploring the Internet's medical-information resources. He had to give step-by-step instructions, because at the time he wrote the book, many materials were still in text-only form, accessible only through a series of typed commands, some downloadable only through a multistep file transfer process. However, Ferguson insisted the information was worth the effort. Resources included specialized e-mail lists and newsgroups such as alt.support.cancer, as well as proprietary content provided to members by the competing AOL, Prodigy, and CompuServe Internet service providers. Just emerging was the World Wide Web, which Ferguson found easily usable and graphically "dazzling," although initially thin on content and sadly lacking in discussion groups (239).

However, as Ferguson predicted, within a few years the best information resources migrated to the Web and new ones proliferated. Huge numbers of medically oriented Web sites appeared, hosted by government research centers, universities, hospitals, nonprofits, commercial interests, and individuals. One of the earliest and largest sites for cancer, Oncolink at the University of Pennsylvania, logged more than a quarter of a million visitors monthly. Much larger was the National Library of Medicine site, www.nlm.nih.gov, which provided a free gateway to the contents of forty-five hundred medical journals (also available in Spanish as of September 2002), a constantly updated listing of clinical trials, a toxicology library, databases of genetic information, information on environmental pollutants, AIDS, bioethics, and a directory of health organizations, among its many features.

At the same time, commercial enterprises such as WebMD, drkoop.com, and Gazoondheit provided user-friendly articles and numerous discussion groups. These sites garnered billions of dollars' worth of market capitalization when they went public in the late 1990s. At its height, drkoop.com, founded by the former U.S. surgeon general, logged 5.5 million visits a month, although a hoped-for advertising bonanza did not materialize for any of the commercial sites (Sheff 2000, 91).

Individuals with a medical message to share with the world used increasingly easy-to-use software to produce their own sites; for example, the Harvard professor and LSD promoter Timothy Leary, suffering from prostate cancer, detailed his day-to-day coping strategies (including his recipe for marijuana-spiked cheese and crackers) until his death in 1996 and, in fact, considered videocasting his own death. At the other extreme, Quackwatch was a Pennsylvania doctor's project to debunk nontraditional medical treatments. Some lay people created Web sites about their own medical treatments to help others, in the spirit of the Internet's renowned "gift culture."[1]

Various studies have confirmed the popularity of this cornucopia of free information. The Science Panel on Interactive Communication and Health cited four studies to show that patients want health information, access to it leads to better medical results, and being informed tends to enhance the doctor-patient relationship (Eng and Gustafson 1999, 32). A 2000 Poynter Internet study showed that more Americans were using the Web to find medical information than for most other purposes, including online shopping (Fox and Rainie). A 2002 Harris poll showed that a growing number of Americans and also many French, Germans, and Japanese routinely sought out and trusted online health information and believed it would improve their relationships with doctors (Taylor and Leitman). The poll found that about 110 million Americans look for health information online an average of three times a month. In Great Britain, the situation was similar: The National Health Service consumer site, for example, had 171,900 visitors in December 2001 (Powell 2002).

Medical material may not go through a centralized approval process. Several studies have found that even professionally produced Web content can be unsound; for example, pediatricians at Ohio State University compared American Academy of Pediatrics data on childhood diarrhea against information found on sixty Web sites sponsored by medical professionals. They found that all but twelve sites had errors;

some of the information was "just garbage" one pediatrician told an interviewer (Kiernan 1998). Canadian researchers found forty-seven different rating instruments that attempted to set standards for online medical content, leaving the authors to wonder whether any standard setting would be possible or even desirable (Jadad and Gagliardi 1998). In 1998, the Clinton Administration proposed legal action to protect consumers from erroneous Web content, but the effort proved fruitless (Kettner 1998). As a result, lay people gathering medical information that may literally be a matter of life and death to them have only the content provider's reputation and their own critical thinking to help them judge its quality.

From the online ferment of the late 1990s emerged one medical communication form that seemed quite satisfying and credible to those who engaged in it—peer communication for mutual help. Thanks to the ease of shareware bulletin-board technology, it was not much of a project for a volunteer to launch a forum. Some of those simple sites became part of the support mechanism for thousands of seriously ill people. Ferguson, who from his early advocacy went on to produce a quarterly newsletter, surveyed 1,000 regular users of online discussion groups and found that a majority of the 191 respondents preferred these groups to their own doctors in the following categories: convenience, cost effectiveness, emotional support, compassion/empathy, help in dealing with death and dying, medical referrals, coping tips, in-depth information, and "most likely to be there for me in the long run." They preferred doctors only for diagnosis and treatment (Ferguson and Kelly 1999, 1). Although Ferguson and Kelly had surveyed a self-selected niche audience, it is nevertheless notable that 76 percent of respondents said online support communities were better sources than doctors for "in-depth medical information," which, as we have seen, is unregulated. How do these online discussions work, and what are the implications for journalism? To address this question, I use data from a multimodal case study that took place from 1998 to 2001.[2]

Peer Communication among Breast Cancer Survivors

It was February 2001—the end of summer in New Zealand, where Reigan Allan, a witty twenty-two-year-old university student and fashion model fighting advanced breast cancer was applying for inclusion in a clinical trial because she was losing ground against conventional remedies. At the same time, she was curious about a new American drug, Herceptin, that she had seen mentioned in discussions on the BCANS forum. Unfortunately, the New Zealand government would not pay for the $10,000-per-dose treatments.

Links to news articles and to the manufacturer's Web site helped Reigan to determine that Herceptin might be her best hope. After the clinical trial failed to help, her family decided to pay for Herceptin treatments. The results were dramatic—Reigan posted that she was no longer confined to her bed, but was starting graduate school and seeing her friends again. Her remission was featured in a TV news report and a magazine cover story (Moore 2001). Her online friends sent messages to the New Zealand health minister, pleading for the government to cover Herceptin treatments for all women with breast cancer, a campaign that eventually succeeded. Meanwhile, Reigan's teddy bear collection overflowed as furry gifts began to arrive from well-wishers around the globe.

During the same month, the BCANS site was also busy with these postings:

- In Switzerland, a young mother of two was fearfully awaiting the results of a bone scan.
- An American sent news of new pain-control techniques and posted a link to a relevant article in *Business Week*.
- From snowbound Nova Scotia came a cheerful post from a nurse just out of the hospital after a four-day stay for a high fever. She had her sights firmly fixed on a May vacation with her husband and toddler. Elsewhere in the province, someone was assembling a team of breast cancer survivors for a distributed-computing project that would use their excess computer processing power to crunch cancer data.
- A Manitoba woman posted that she and her husband were enjoying their Mexican vacation—first break in a long time without their daughters—except for nagging stomach pain that might be her gallbladder. Sadly, she was to return home to a horrifying test report: The pain was being caused by widespread liver metastases. She died two months later.

• A physician, who had been watching over the Web site for three years to help with technical questions, weighed in with information on side effects of the commonly prescribed, but controversial, drug tamoxifen. From Florida, someone chimed in with a hyperlink to an online interview with a scientist, "the Father of tamoxifen." More than two hundred women had already filled out a Web site–sponsored survey about their own widespread problems with tamoxifen.

At this time, BCANS was receiving about 175 messages daily and hosting five times as many visits for a total of some thirty thousand visits a month. Incoming messages were not filtered by a gatekeeper, organized hierarchically, or subjected to the other sense-making techniques by which print journalists assist the reader. However, the Webmistress in fact used more organizing systems than any newspaper or TV station to sort the content and provide for flexibility of use. She used shareware message-board technology that allows for threaded discussions, giving participants the option of posting a message under an ongoing topic heading or else starting a new topic thread under a new heading. Threaded discussions allow each contribution to be visually related to all of the others in a sort of braid of giving, a "chain of all things good," as one woman put it. Three different message areas ("boards") allowed users to presort their postings according to whether they were about medical issues, "off-topic" (jokes, recipes, general chitchat), or get-together planning. All messages—more than a quarter of a million of them—were saved in a searchable archive, so a visitor could, for example, use the keyword "Xeloda" to pull up all messages on that drug.

The site also had separate areas containing a glossary, surveys, research papers, book recommendations, and autobiographical sketches of more than three hundred participants. It even had a page that automatically displayed the daily breast cancer news from Yahoo! and CNN. A special chat room served the desires of those who craved some real-time group conversation across the miles. The site never slept, with visitors from every continent logging in around the clock, although Americans were present in the greatest numbers (79%), with Canadians composing the second-largest group.[3] In a typical comment, one woman wrote (Harbauer 1999, 10): "I look forward to my early mornings with you. . . . Instead of rising early to befuddlement, I have found a little more clarity." In January 1999, when the site went offline for three days because of heavy use, the Webmistress was besieged with e-mails from women frantically asking where to send their "fees" so their "memberships" would not be discontinued.

The online discussion was ever changing; for example, there were 468 different thread topics during January 2001 in the main discussion forum, with an average of 11 responses to each topic. The number of responses varied from 0 to 111. A few informational postings did not require a response and received none, such as "Zoledronate trial info," "Vitamin E and Radiation," and "Attn: Atlantic Canadian Women: Atlantic Breast Cancer Information Kit Evaluation." At the other extreme were outpourings of support for long-time, active forum members: "Bus Ride to Doctor's Appointment" (111), "Results of Meeting" (85), and "Rally for Our Friend" (49). Other long threads came in response to "How Not to Curse Yourself" (43), "Sex Drive" (31), "Radiation???????" (41), and "Herbs: For Your Consideration" (40). Medical queries ranged widely, from "Fatigue? Any Ideas?" to "Tumor Markers???" to "Taxol vs. Taxotere" to "Hair Loss and Wig Liners." Responses were softened with caring comments such as, "Let us know how you're doing" and "We're here for you, 24/7." The more personal messages centered on test results, stress, and philosophical insights. A husband reported that his wife was in the hospital, and later in the month posted heartbrokenly of her death. "Onc's Visit Today," "Some Results in, But Still Waiting. . .,"and "Need Prayers" are other examples. Most messages collected a string of responses such as, "It warms my heart to know that you are feeling better and hopefully soon you will be back to doing the things you enjoy."

The forum also was occasionally a venue for advocacy and critique of the medical establishment. One woman declared: "Doctors are used to having the last word. . . . Don't let them intimidate you!!! YOU ARE THE PATIENT." Widely welcomed was a Florida woman's campaign to remind medical personnel to consider each patient as a whole person; she produced and gave away more than two hundred pink buttons reading, "TO YOU, IT'S A JOB; TO ME, IT'S MY LIFE." Several women announced they were discontinuing the tamoxifen regime because of burdensome side effects that had been chronicled in a report by a BCANS member (Radcliffe 1999). One wrote that she told her doctor she was going off raloxifene because of bone pain. Although he was "stern" and unwilling at the start of the conversation, he soon assented, she reported. A woman who undertook to be more assertive with her oncologist reported happily, "It didn't seem like the same doctor."

Over the years, many postings referred to the isolation of having breast cancer. Many said it was a relief to find a group of women in similar circumstances. An Australian reflected, "I cannot imagine how isolating it would have been to go through it alone." A number of participants commented that above all, they preferred information from "others who have been there."

Rheingold noticed that people engaged in ongoing discussions like the one just described care about each other, take on various roles within the group, and form a cohesive online community with a "grassroots groupmind" that functions as "a merger of knowledge capital, social capital, and communion" (1993, 110). This "groupmind" quality gives people a way to be with others who share their concerns.

At first, it was not at all evident to health-care authorities that communication among peers had any value or was even advisable in the absence of expert guidance. Although the potential of computers for health communication was generally appreciated, emphasis initially was on the "transmission model" aimed at educating patients rather than having them help each other. An early experiment in this direction was the Comprehensive Health Enhancement Support System (CHESS) designed at the University of Wisconsin, Madison (Hawkins et al. 1997). This successful top-down system had six modules, including one for breast cancer. It included a frequently asked questions (FAQ) section; an archive of articles from journals, brochures, and the popular press; a section on how to obtain help and support; a dictionary; an ask-an-expert service; self-assessment questions; personal stories; and online health charts and profiles. What it did not have was a way for patients to construct their own medical narratives.

However, health care was changing, the paternalistic model giving way to a reconceptualizing of the patient as an active, critical "health-care consumer" and the doctor as a "health-care provider" (see, for example, Porter-O'Grady and Wilson 1995; Roter and Hall 1997; Eng and Gustafson 1999). Coincidentally (or not), in online medical communities, the user also takes a nontraditionally active role by providing some of the content. Early research into this phenomenon suggested that "similar circumstances make personal stories more relevant, information more credible, and empathy more likely" (Scheerhorn 1997, 175). In addition, he said, participants can help each other track down good doctors and cheaper drugs, treatments, and equipment. Brennan and Fink envisioned several advantages to online peer support: Social anonymity and asynchronous communication lead to a feeling of "less inhibited and more assertive communication . . . more equal participation, more uninhibited verbal behavior, and greater choice shift" (1997, 166). Brennan and Fink observe that conversing online is more convenient, and the option of remaining hidden may reduce stigma and embarrassment. This protection can lead to brave and helpful comments like the following posts on death that appeared on the BCANS site:

> Since I have finished all my treatments I feel as though I am waiting for the other shoe to drop. I sometimes wake up in the middle of the night thinking I must do something else. I have never told anyone that before.

> When I was first diagnosed I thought of death constantly. Didn't even want to buy new clothes, because I didn't feel I would be around to wear them. We had to move and get new furniture and I kept thinking, "I hope Steve's second wife likes this stuff." We joined a memorial society that has several funeral homes that offer reasonable rates. Mostly I worry that if I get a recurrence and don't die immediately, it will ruin us financially. [People] do everything but run screaming from the room when I bring [death] up, so I let them go on with their pathetic shallow lives.

These varied comments give any reader rich, comforting, even humorous perspectives on a much-avoided topic. This illustrates one of a Web discussion's great strengths—its ability to gather the wisdom of many people for the benefit of any who might choose to peruse it.

Like many stable online communities, this group also went offline sometimes to exchange visits, phone calls, cards, and gifts. In addition, several dozen members collaborated on three books. *Breast Cancer Online: In Our Own Words* (downloadable in PDF format from the Web site) was created from 1998 to 1999 when eleven women volunteered to sort through thousands of postings on topics like "Surviving Cancer and the Holidays," "What a Friend Should Say," and "Am I Crazy to Have This Feeling?" Women contributed recipes and artwork for *Penney's Friends Cookbook*, a tribute to a popular thirty-four-year-old Minnesota group member who had recently died (Harbauer 1999). Most recently, *In Our Own Words How We Told Our Children* was published in 2001 as the second in the Breast Cancer Online Series. All of these projects placed participants squarely in the role of author, and in fact several who worked on the publications were professional editors. Thus, there was much overlapping of roles and blurring of boundaries in true postmodern fashion.

Conclusion: "Groupmind" Versus Journalism

Online communication reconfigures both audience and content. In this regard, it is no different from other communication technologies, from smoke signals to telephones, printing press to silver screen. Emergent technology influences the form and content of expression, giving rise to new content providers, meeting a society's needs and desires in new ways, sometimes even tilting the balance of political power. The particular qualities of digital communication, and especially of the easy-to-use World Wide Web, allow certain medical-communication needs to be addressed as never before. Global accessibility promotes formation of niche groups of patients who, among themselves, have vast stores of experience, empathy, and skills in particular areas as well as motivation to use these for the common good—forming a powerful "gift culture" or, as Rheingold recognizes, generating copious social capital. People whose knowledge may be devalued in traditional medical settings here create their own frameworks in which the knowledge is welcomed, built on, critiqued, and archived to form a growing body of searchable material. Sometimes their writing is even selected for inclusion in a book. In true postmodern fashion, the traditional roles of expert and lay person, author and audience, begin to merge.

In short, peers in online medical-communication situations shape their discourse to suit themselves, and that discourse is different from anything the mass media can provide. Primarily, it simply reiterates prevailing medical wisdom, but it fills in the small details with personal anecdotes, drills down for the latest professional studies, weaves in the day's news, and ties up the information package with a bow of personal respect and support.

The traditional mass media know from decades of study that audiences are interested in medical issues, and they have tried to address this demand with special reports, weekly news sections, and TV shows. In recent years they have used company Web sites to supplement their reports with details that might otherwise have been discarded due to limited airtime or print space. However, these approaches may not be the best way for the mass media to adjust to digital medical communication. This is because the "grassroots groupmind" approach of online medical communication does not compete with mass communication so much as complement it. Although the Web is considered a many-to-many or even a mass communication channel, which might theoretically make it a competitor of traditional journalism, in fact the appeal of peer online communication is its interpersonal side—the empathy, the personal advice, and the relationships it offers. News and information are just a click away, but they are in the background unless someone decides to click. A mass medium cannot meet the ravenous needs of patients for this personal support and niche information.

Since the Web is very successfully addressing the needs of a significant and growing number of patients, the path for traditional journalists is clear: Instead of trying to fragment their coverage further in attempts to engage an increasingly fragmented audience, journalists should go the opposite way. They should provide more comprehensive, useful, patient-centered coverage of the larger picture, the issues that ultimately affect everyone in the medical realm; for example:

- When reporting on political budget maneuvers, write a sidebar on some of the medical programs that are at stake and explore the implications. Report on the politicians and special interests trying to influence decisions, making sure to interview patient advocates and look independently at the facts.
- Analyze available epidemiological statistics, searching for trends in hospital mortality rates, drug affordability, research failures and successes, and other issues. Use this information to put medical news in perspective.

- Keep the patient in mind when reporting on announcements of breakthroughs, new treatments, and new technologies. For example, answer questions like, When and where will this new treatment be realistically available to most of us? Does this mean that ever fewer of us will have access to ever more sophisticated (or at least costly) treatments? Which patients have the best chance of benefiting from this? How long might it take for negative side effects to show up; will people who use this new product be offering themselves up as guinea pigs? Aggressively follow up later to see what really happened.
- Bring alternative perspectives into medical reporting without trivializing foreign health-care systems or unfamiliar remedies.
- Independently explore connections that have been obscured by special interests, such as links between pollution and cancer, chemicals and immune-system disorders, poverty and ill health. Examine why drug companies turning billion-dollar profits are charging so much for their products. Challenge the use of public-relations euphemisms such as "air quality" when "air pollution" is really meant.
- Above all, track important policy making in clear, understandable ways. Give information that allows patients to make a difference before decisions are made, such as bill numbers, hearing schedules, and addresses to submit comments.

Helping one individual at a time is what the "grassroots groupmind" of online medical communication does best. The complementary strength of journalism, as it should be practiced, is in delivering information that empowers a mass audience to steer its own society. Few issues are so close to basic human rights as health care; trained journalists understand the legislative process that determines the current situation, and through their position they have special access to relevant individuals and data. By wielding this power responsibly and effectively, traditional journalists can complement online medical communication and appeal to today's more critical, active patients. Working together, these potentially complementary styles of medical communication can promote improvements in health care at a time when they are urgently needed.

Notes

1. A "gift culture," such as the potlatch culture of the Pacific Northwest nations, confers status according to the value of one's publicly bestowed gifts. In *The Cathedral and the Bazaar* (1999), Eric S. Raymond frames the Linux open-source software movement as emerging from a "bazaar" of free programmer contributions, contrasted with a monolithic corporate "cathedral" that gives nothing away. The "gift culture" metaphor has frequently been applied to the otherwise puzzling outpouring of personal and institutional effort on the Web.

2. This multimodal case study, conducted as a dissertation research project at the University of Washington, is posted in a set of PDF files at http://bca.ns.ca/pat.

3. Although a majority of site users were women, men also (rarely) get breast cancer, and a few of these survivors also participated. A sprinkling of participants included the spouses, siblings, children, or friends of women with breast cancer.

Bibliography

Batt, Sharon. *Patient No More: The Politics of Breast Cancer*. Montreal: Gynergy/Ragweed, 1994.

Becker, Shawn H., and Stephen J. McPhee. Health-Care Professionals' Use of Cancer-Related Patient Education Materials: A Pilot Study. *Journal of Cancer Education* 8(1) (1993): 43–46.

Bilodeau, Barbara A., and Lesley F. Degner. "Information Needs, Sources of Information, and Decisional Roles in Women with Breast Cancer." *Oncology Nursing Forum* 23 (1996): 691–696.

Brennan, Patricia F., and Sue V. Fink. "Health Promotion, Social Support and Social Networks." In *Health Promotion and Interactive Technology: Theoretical Applications and Future Directions*, edited by Robert L. Street Jr., William R. Gold, and Timothy Manning, 157–169. Mahwah, N.J.: Lawrence Erlbaum, 1997.

Butow, Phyllis N., et al. "When the Diagnosis Is Cancer: Patient Communication Experiences and Preferences." *Cancer* 77(12) (15 June 1996): 2630–2637.

Colomeda, Lorelei A. L. *Through the Northern Looking Glass: Breast Cancer Stories Told by Northern Native Women*. New York: National League for Nursing, 1996.

Cooper, Cary L., ed. *Stress and Breast Cancer*. Chichester, N.Y.: John Wiley & Sons, 1988.

Dance, Betsy. *First Aid Yourself: Essential Breast Cancer Websites*. Manakin-Sabot, Va.: Hope Springs, 2000.

Deane, Karen A., and Lesley F. Degner. "Information Needs, Uncertainty, and Anxiety in Women Who Had a Breast Biopsy with Benign Outcome." *Cancer Nursing* 21 (1998): 117–126.

Degner, Lesley F., et al. "Information Needs and Decisional Preferences in Women with Breast Cancer." *JAMA* 277(18) (14 May 1997): 1485–1492.

DeGrasse, Catherine E., Kylie Hugo, and Ronald C. Plotnikoff. "Supporting Women during Breast Diagnostics." *The Canadian Nurse* (October 1997): 24–30.

Eng, Thomas R., and David H. Gustafson, eds. *Wired for Health and Well-Being: The Emergence of Interactive Health Communication*. Washington, D.C.: U.S. Department of Health and Human Services, U.S. Government Printing Office, 1999.

Ferguson, Tom. *Health Online*. Reading, Mass: Addison-Wesley, 1996.

Ferguson, Tom, and William J. Kelly. "E-Patients Prefer E-Groups to Doctors for Ten out of Twelve Aspects of Health Care." *The Ferguson Report* 1(1) (January–February 1999): 1–3.

Fox, Susannah, and Lee Rainie. "The Online Health Care Revolution: How the Web Helps Americans Take Better Care of Themselves." Washington, D.C.: The Pew Internet and American Life Project, 2000, at www.pweinternet.org.

Gee, Elizabeth D. *The Light around the Dark*. New York: National League for Nursing, 1992.

Girgis, A., R. W. Sanson-Fisher, and W. H. McCarthy. "Communicating with Patients: Surgeons' Perceptions of Their Skill and Need for Training." *Australian and New Zealand Journal of Surgery* 67(1) (November 1997): 775–780.

Glanz, Karen, Frances Marcus Lewis, and Barbara K. Rimer, eds. *Health Behavior and Health Education: Theory, Research, and Practice*. San Francisco: Jossey-Bass, 1997.

Haines, Judith. "Breast Cancer Resources Online." *The Canadian Nurse* 93(1) (1997): 49–50.

Harbauer, Julie. *Penney's Friends Cookbook*. Springfield, Ill.: Self-published, 1999.

Hawkins, Robert P., et al. "Aiding Those Facing Health Crises: The Experience of the CHESS Project." In *Health Promotion and Interactive Technology: Theoretical Applications and Future Directions*, edited by Richard L. Street Jr., William R. Gold, and Timothy Manning, 79–102. Mahwah, N.J.: Lawrence Erlbaum, 1997.

Heaney, Catherine A., and Barbara A. Israel, "Social Networks and Social Support." In *Health Behavior and Health Education*, edited by Karen Glanz,

Frances Marcus Lewis, and Barbara K. Rimer, 179–205. San Francisco: Jossey-Bass, 1997.

Jadad, Alejandro R., and Anna Gagliardi. "Rating Health Information on the Internet: Navigating to Knowledge or to Babel?" *JAMA* 279(8) (25 February 1998): 611–614.

Johnson, J. David. *Cancer-Related Information Seeking.* Cresskill, N.J.: Hampton, 1997.

Kettner, Kristin B. "Networked Health Information: Assuring Quality Control on the Internet." *Federal Communications Law Journal* 50(2) (March 1998): 417–439.

Kiernan, Vincent. "Study Finds Errors in Medical Information Available on the Web." *Chronicle of Higher Education* 21 (June 1998): 20–21.

Klemm, Paula., K. Repper, and L. Visich. "A Nontraditional Cancer Support Group: The Internet." *Computers in Nursing* 16(1) (1998): 31–36.

Kurtz, Linda F. *Self-Help and Support Groups: A Handbook for Practitioners.* Thousand Oaks, Calif.: Sage, 1997.

Kushner, Rose. *Breast Cancer: A Personal History and an Investigative Report.* New York: Harcourt Brace Jovanovich, 1975.

Lerman, C., et al. "Communication between Patients with Breast Cancer and Health Care Providers: Determinants and Implications." *Cancer* 72(9) (1 November 1993): 2612–2620.

Lorde, Audre. *The Cancer Journals.* San Francisco: Aunt Lute Books, 1980.

Luker, Karen A., et al. "The Information Needs of Women Newly Diagnosed with Breast Cancer." *Journal of Advanced Nursing* 22(1) (July 1995): 134–141.

Luker, Karen A., et al. "Information Needs and Sources of Information for Women with Breast Cancer: A Follow-up Study." *Journal of Advanced Nursing* 23(3) (March 1996): 487–495.

Middlebrook, Christina. *Seeing the Crab: A Memoir of Dying Before I Do.* New York: Doubleday, 1996.

Moch, Sandra D. *Breast Cancer: Twenty Women's Stories.* New York: National League for Nursing, 1995.

Moore, Jenna. "The Brave and the Beautiful." *Next* 208 (November 2001): 76–81.

Nielsen, Jerri. *Icebound: A Doctor's Incredible Battle for Survival at the South Pole.* New York: Hyperion, 2001.

Porter-O'Grady, Tim, and Cathleen K. Wilson. *The Leadership Revolution in Health Care.* Gaithersburg, Md.: Aspen, 1995.

Powell, John. "The WWW of the World Wide Web: Who, What, and Why?" *Journal of Medical Internet Research* 4(1) (2002): E4.

Radcliffe, Nancy. "Coping with the Cure," 1999, at http:/bca.ns.ca/surveys.

Ray, Colette, and Michael Baum. *Psychological Aspects of Early Breast Cancer.* New York: Springer-Verlag, 1985.

Raymond, Eric S. *The Cathedral and the Bazaar: Musings on Linux and Open Source by an Accidental Revolutionary.* Petaluma, Calif.: O'Reilly, 1999.

Reeve, Jean, and Lynne Wagner, eds. *Breast Cancer Online: In Our Own Words.* Halifax: Breast Cancer Action Nova Scotia, 1999, at http://bca.ns.ca/booklets/inourown.pdf.

Rheingold, Howard. *The Virtual Community: Homesteading on the Electronic Frontier.* Reading, Mass.: Addison-Wesley, 1993.

Rice, Ronald E., and James E. Katz, eds. *The Internet and Health Communication: Experiences and Expectations.* London: Sage, 2001.

Roter, Debra L., and Judith A. Hall. "Patient-Provider Communication." In *Health Behavior and Health Education: Theory, Research, and Practice,* edited by Karen Glanz, Frances Marcus Lewis, and Barbara K. Rimer, 206–226. San Francisco: Jossey-Bass, 1997.

Scheerhorn, Dirk. "Creating Illness-Related Communities in Cyberspace." In *Health Promotion and Interactive Technology: Theoretical Applications and Future Directions,* edited by Robert L. Street Jr., William R. Gold, and Timothy Manning, 71–85. Mahwah, N.J.: Lawrence Erlbaum, 1997.

Sharf, Barbara F. "Communicating Breast Cancer On-Line: Support and Empowerment on the Internet. *Women & Health* 26(1) (1997): 65–84.

Sheff, David. "Net Physician." *Yahoo Internet Life* (February 2000) 91–95.

Spiegel, David. "Health Caring: Psychosocial Support for Patients with Cancer." *Cancer* 74(4, supplement) (15 August 1994): 1453–1457.

———. "Psychosocial Aspects of Breast Cancer Treatment." *Seminars in Oncology* 23(1, supplement 1) (February 1997): S1-36–S1-47.

Spiegel, David, et al. "Effects of Psychosocial Treatment on Survival of Patients with Metastatic Breast Cancer." *The Lancet* 2(8668) (14 October 1989): 888–891.

Street, Richard L. Jr., William R. Gold, and Timothy Manning, eds. *Health Promotion and Interactive Technology: Theoretical Applications and Future Directions.* Mahwah, N.J.: Lawrence Erlbaum, 1997.

Street, Richard L. Jr., and B. Voigt. "Patient Participation in Deciding Breast Cancer Treatment and Subsequent Quality of Life." *Medical Decision-Making* 17(3) (1997): 298–306.

Taylor, Humphrey, and Robert Leitman. "4-Country Survey Finds Most Cyberchondriacs Believe Online Health Care Information Is Trustworthy, Easy to Find and Understand." *Health Care News* 2(12) (11 June 2002): 1–3.

Thompson, Barbara (ed.). *Breast Cancer Online: How We Told Our Children.* Dartmouth, N.S.: Print Atlantic, 2001, at http://bca.ns.ca/booklets/howwetold.pdf.

Wilson-Hashiguchi, Clo. *Stealing the Dragon's Fire: A Personal Guide and Handbook for Dealing with Breast Cancer.* Bothell, Wash.: Wilson, 1995.

Conclusion

Kevin Kawamoto

THIS BOOK IS NOT INTENDED to be the authoritative source on digital journalism but to help illuminate the concept—offering explanations and ideas of where it came from, what it is, and where it is going—and advance the discussion of this topic with journalism students, educators, media professionals, and others. We hope that by now you have a sense of the breadth and complexity of digital journalism, which could be a product (such as a news Web site), a process or practice (the work and activities of the journalist in an online environment), and even a philosophy (the values and beliefs associated with gathering and presenting news using digital technologies).

The appeal of digital journalism in journalism and communication programs across the country is undeniable. The flashiness—sometimes literally—of technology can seem hip and cool to those contemplating a future as a communications professional. But as the preceding chapters reveal, digital journalism is serious business. As in previous eras when technology—be it the printing press, telegraph, telephone, wireless communication devices, fax machines, tape recorders, satellites—affected the work of journalists in sometimes revolutionary ways, computers and other digital technologies are changing the face (but not necessarily the soul) of journalism today.

Digital journalism can affect all stages of the storytelling process, from conceptualization to presentation and even beyond as the story generates audience feedback and interaction. A journalist might use e-mail, browse electronic discussion groups, and surf the Web to come up with and develop a story idea. Computer-assisted research could then be used to gather data and conduct an analysis. Although these terms sound technical and scientific, they need not be. Broadly speaking, data can include facts, figures, documents, interviews, maps, graphics, digital photographs, audio files, and other forms of information that have yet to be analyzed, processed, and integrated into a story.

The journalist, perhaps with assistance from a digital design and layout expert at the news organization, can use computer hardware and software to tell the story in a multimedia format. The interface between journalist and public might be a Web site or a handheld multimedia device or, in the future, something we haven't even thought of yet. (Because new ways of delivering and accessing news stories are emerging, news organizations should conduct or support audience-usability research so that contemporary news producers can develop technologies and content most suitable for their intended audiences.)

This is just a thumbnail sketch of one way that digital journalism might manifest itself in practical terms. It should not imply that new practices will necessarily supplant the old, however. The human side of journalism—good interpersonal communication, a commitment to public service, a respect for reliable sources, and a sense of obligation to seek truth with fairness, balance, and context—represents timeless skills and values. Simply put, journalists must continue to be intelligent and ethical people.

Also, the basic goal of effective storytelling—whether a hard news story or a long narrative feature in three parts—remains constant regardless of the technologies used. Digital media, as mentioned elsewhere in this book (but which bears repeating), facilitate functionally and structurally dynamic storytelling. A journalist can tell a more complete story in a digital format than in the space- or time-limited formats of traditional media. Of course, the fact that one can doesn't mean one does, but the potential is present.

A simple example to illustrate this would be a story about violence and conflict in the Middle East. To understand this problem fully in its historical, political, and social context would require considerable airtime and news hole, neither of which traditional media are equipped to provide. Newspapers and magazine articles may be long, but even the lengthiest of these would be bounded by space limitations for such a complex subject. Television news stories tend to be short because so many of them need to be crammed into a limited time frame and also because of long-held journalistic conventions practiced in that medium. News stories can often be measured in seconds, not minutes. With digital media, the inflexible constraints of time and space are less relevant. A multimedia Web site on Middle East problems would likely require the equivalent of thousands of pages of newsprint or magazine pages if it were to discuss current events in their proper context. Covering key events and developments from the present day back to biblical times does not lend itself to brevity. Text, photos, video, sound, graphics, maps, satellite images, slide shows, discussion groups, interactive quizzes, and so forth would all be part of the presentation. Visitors to this site could experience as much or as little of the full story that they wanted. The available content would be enormous, but it would not all have to be digested at once, or even in any particular order.

With hypertext links, interactivity, nonlinear presentation of information, and other features, the limitations shift from the medium to the user. Greater news volume does not necessarily mean greater news consumption. While the medium can accommodate large amounts of content, most people do not have unlimited amounts of time. Thus, the time constraints of the user become an issue in a digital news world. People frequently complain about not having enough time, about feeling rushed or feeling stressed by all that needs to be done in a day with too few hours. Some might even argue that shorter news stories are more practical in a busy world. This solution is short sighted. Not everyone will agree with this statement, but ultimately too much information is far better than too little, just as an overabundance of food is better than a scarcity. This means that the news consumer (or the food consumer) must consume judiciously, as appropriate to his or her particular needs and circumstances. A self-regulated media diet puts control into the hands of the consumer of news, not the producer. A free and democratic society is better off risking information overload than risking information scarcity.

The Stakes Are High

A free press has been called the cornerstone of democracy. There is a story about a journalism professor who used to tell his students on the first day of class, "The work that you do as journalists may determine whether more than two hundred years of democracy stands or falls in the future of this country. But don't let me pressure you." Journalists throughout history have often been agents of social change—sometimes in small ways, other times in quite revolutionary and fundamental ways. They have exposed corruption in government and corporations, shed light on tragedies and disasters (famine, disease, violence) at home and abroad, brought to the public's attention specific and systematic cases of social injustice, reported on conflicts in communities and organizations, publicized the consequences of criminal activity, focused on efforts by people to help improve society and the world, listened to the accusations of the underdog, whistle-blower, and victim for stories of social significance and newsworthiness. By doing these things, they have at times been able to influence public opinion and public behavior. The free press is not merely an abstraction; it is a powerful tool that, when used responsibly, balances and illuminates the contending forces of government, market, and civil society. The free press is not any one thing, but rather a composite of many things at once. It comprises an underlying philosophy and the people, technologies, activities, systems, processes, and motivations that give that philosophy life and meaning.

This underlying philosophy that responsible journalism is essential to perpetuating a free and democratic society is what distinguishes digital journalism from mere content provision. (Journalists should not be content to be called "content providers"!) This may sound like a lofty enterprise, and certainly the reality of day-to-day reporting on routine topics may make the profession feel more like a job than an exercise in democracy building, but in the bigger picture, tyranny and ignorance thrive without a free press. Digital journalism, like any other form of journalism, plays a role in making sure that they do not. With the burden and privilege of this responsibility in mind, journalism students should learn to exploit the power of digital media.

Browse, Observe, and Analyze

Serious students of digital journalism have a tough job ahead of them. The field of study is a moving target in that it keeps changing. The digital journalism of the 1980s is different from the digital journalism of the early 2000s. Inevitably, it will continue to evolve, as must its practitioners.

Becoming literate and proficient in digital journalism requires more than reading a book. Like surgery or sports, you have to "do" it to get good at it. Journalism courses that offer a hands-on component with digital-media technologies are one way to get started. These courses may be called digital journalism, cyberjournalism, multimedia journalism, new media journalism, Web journalism, Internet journalism, interactive journalism, online journalism, and computer-assisted journalism (although this last one may be geared more toward social science research using computers). The time is approaching when it won't be necessary to label digital journalism courses with such monikers. The digital component will be integrated into ordinary journalism classes. An advanced news-writing course, for example, will have a digital component built into the curriculum. Perhaps at some schools that time has already arrived.

Internships are another way to learn the ropes and have long been an important component of journalism education. They are not always easy to acquire, however. Learning on the job—also known as experiential learning—has proven highly effective under the right conditions. Unfortunately, internships can also be exploitive under the wrong conditions. Many internship sites do not pay interns, which is not necessarily bad. What the student should get in return for volunteering at an internship site is the opportunity to learn new skill sets and knowledge, but this does not always happen. An internship is dysfunctional when the intern is asked to perform exclusively menial jobs that have nothing to do with journalism. The intern, then, simply becomes a source of free labor. This is not to say that the intern should never do any kind of menial work; sometimes these things are just a realistic part of the work culture. But those tasks should be balanced with some meaningful learning experiences as well.

A faculty or staff member who is in contact with a site supervisor at the internship location should oversee all mutually productive internships. Internship responsibilities should be clearly spelled out in a student contract. A description of the work to be performed by the intern along with a list of learning goals should be part of this contract, in addition to information about expected work hours and days. If student interns are getting academic credit or a grade (or both), they should know exactly what they need to do to satisfy the requirements. If there is a problem, there should be a mechanism by which the student intern, site supervisor, and faculty advisor can work together to resolve it. Student interns should be held accountable as well for inadequate internship performance. That is why a clear contract of expectations is essential. Journalism schools usually maintain a list of available internships, but students should also inquire about opportunities that have not been publicized. It is possible that an organization never considered having an intern on board and could be sold on the idea by the right person. Internships sometimes lead to jobs at the internship site, but they often do not, regardless of how well the intern performs. The experience gained could lead to a job elsewhere, however.

Another important way to become literate and proficient in the field is to browse, observe, and analyze what is already out there in terms of digital journalism products. It is imperative to view a large number of Web sites to get a sense of the design and layout possibilities, the range of interfaces used and their effectiveness, the features and benefits of one site versus another, the technique of the multimedia storytelling, the opportunities for feedback and interaction, and other things. Be the digital anthropologist as you surf from one site to another. Take field notes. Consider what works for you, what doesn't, and why.

Online News-Source Links

The best way to start looking at many different news-related Web sites in one extended sitting is to visit a portal-like site that provides hypertext links to hundreds, if not thousands, of news-related Web sites. Below is a list of Web sites that do just that. Although newspaper Web sites are popular sources of local online news, be sure to check out magazine, radio, and TV station Web sites as well. There are also news sources that are found only online and do not have a traditional media partner.

NewsLink.org (http://newslink.org)
Newspaper Association of America (NAA)'s Newspaper Links
(www.newspaperlinks.com/home.cfm)
Kidon Media Link (www.kidon.com/media-link/index.shtml)
Mondo Times (www.mondotimes.com)
ABYZ News Links (www.abyznewslinks.com)

Newspapers

Some of the more popular newspaper Web sites as featured on NewsLink include the following:

Washington Post (www.washingtonpost.com)
Los Angeles Times (www.latimes.com)
New York Times (www.nytimes.com)
Miami Herald (www.miamiherald.com)
USA Today (www.usatoday.com)
New York Post (www.nypost.com)
New York Daily News (www.nydailynews.com)
Atlanta Journal-Constitution (www.accessatlanta.com)
Dallas Morning News (www.dallasnews.com)
Washington Times (www.washingtontimes.com)
Philadelphia Inquirer (www.philadelphiainquirer.com)
Boston Globe (www.bostonglobe.com)
Chicago Tribune (www.chicagotribune.com)
Detroit Free Press (www.detroitfreepress.com)
Phoenix Arizona Republic (www.azcentral.com)
San Francisco Chronicle (www.sfchronicle.com)
Tampa Tribune (www.tampatribune.com)
Orlando Sentinel (www.orlandosentinel.com)
Baltimore Sun (www.baltimoresun.com)
Charlotte Observer (www.charlotteobserver.com)
Chicago Sun-Times (www.suntimes.com)
Cleveland Plain Dealer (www.cleveland.com)
St. Louis Post-Dispatch (www.postdispatch.com)
Indianapolis Star (www.indianapolisstar.com)
Fort Lauderdale Sun-Sentinel (www.sun-sentinel.com)

Television

Here are some examples of popular news-oriented television Web sites:

MSNBC (www.msnbc.com)
ABC News (www.abcnews.com)
CBS News (www.cbsnews.com)
CNN Interactive (www.cnn.com)
Fox News (www.foxnews.com)
PBS (www.pbs.org)

Other Sites

To keep up with research, commentary, and developments in digital journalism, a number of online journals or Web sites should be regularly consulted, including the following:

Broadcasting and Cable (www.broadcastingandcable.com)
Columbia Journalism Review (www.cjr.org)
Editor and Publisher (www.editorandpublisher.com)
Journal of Electronic Publishing (www.press.umich.edu/jep)
Online Journalism Review (www.ojr.org/ojr/page_one/index.php)
The Poynter Institute (www.poynter.org)
(Note: The Poynter Institute's Web site, Poynteronline, has a particularly useful "Journalism Links" page that lists hundreds of links to online publications and programs, news sites, news and journalism resources, journalism organizations, centers and schools, foundations, and journalism jobs. It is one of the most comprehensive collections of journalism-related links on the Web.)
CNN Interactive's "The Anatomy of a Story" (www.cnn.com/ EVENTS/1996/anniversary/how.things.work/index2.html)
(Note: Visit this site for an in-depth explanation of how CNN Interactive packages a news story for its cable TV and online news components. This feature has been around for many years now, but it still provides a useful glimpse into the behind-the-scenes process of online news production.)

Examples of Alternative Media

The Web is an ideal place for so-called alternative publications to flourish. They often provide a perspective that is different from the mainstream "big media," and they are not afraid to show their biases or take advocacy positions. The alternative media have always been an important part of the American media landscape, but the Web allows more of them to have a public presence. The following is just a small sampling of alternative news sources:

AdbustersMonthly (www.adbusters.org)
AlterNet.org (www.alternet.org)
CorpWatch (www.corpwatch.org)
Guerilla News Network (www.guerrillanews.com)
Independent Media Center (www.indymedia.org)
OneWorldGlobal (www.oneworld.net)
Opensecrets.org (www.opensecrets.org)
TomPaine.com (www.tompaine.com)
Undercurrents (www.undercurrents.org)
Working For Change (www.workingforchange.com)

While alternative news sources generally do not have the large unique visitor counts of more mainstream news sites, they do often have a loyal and passionate following, especially among people who, for whatever reason, are not satisfied with mainstream media. Online alternative news sources serve as a virtual watering hole for people who disagree with or dissent from more centrist and mainstream perspectives.

Staying current with developments in digital journalism can also be achieved by regularly doing a search on Google, Yahoo!, or other search engines or portals. Searching the Web using keywords such as "digital journalism," "cyberjournalism," "multimedia journalism," and other descriptors mentioned above will yield a wide range of resources, from conference papers to class syllabi. As you discover informative Web sites that discuss this topic in a way that is helpful to you, make them a part of your regular news menu. Bookmark them on your Web browser or at least write them down somewhere.

Final Thoughts

Digital journalism allows journalists to tell a more dynamic and robust story with fewer constraints than traditional news media. It can enhance a story's depth and breadth and use more ways of engaging the news consumer. Digital journalism, as some of the chapter authors have suggested, can also be empowering for disenfranchised groups, facilitating the flow of news and information among citizens and activists. At the same time, it can also blur the distinction between a professional journalist and a nonprofessional. Some people may think this is a good thing; others are more wary. Does the ease of access to a publishing medium like the Web compromise accuracy, balance, fairness, and other principles of ethical writing? Or does this access level the playing field—allowing a greater diversity of voices to be heard? These are loaded questions, meant to provoke discussion. The answer to each of these questions could be both yes and no.

Digital journalism also has implications for privacy, manipulation of digital photographs or video, the separation between editorial and business functions of a media organization, the development of grassroots channels of communication, democracy, and civic engagement, news coverage during times of crisis, and more. At a time when the news media have suffered from a lack of trust and credibility with the public, it is of utmost importance that those who study and work in digital journalism also be concerned with ethics and professional standards. While it is true that responsible journalism can strengthen a democracy, it is probably also true that irresponsible journalism can weaken it. A society that distrusts its journalists is in trouble.

A Final Note to Students

Digital journalism has become an important part of the media landscape and will continue to evolve and mature in the years ahead. You are in a position to contribute to that evolution, if not influence it to some degree. It is hoped that these views from the horizon, written by people who know and understand the field of digital journalism, have given you some ideas about the potential of emerging digital-media technologies and the role you could play as a media professional in these newsworthy and vital times. Good luck!

Multimedia Coverage:
A Case Study and Exercise

Multimedia News Package

NEWS ORGANIZATIONS ACROSS THE COUNTRY went into high gear after a fire killed ninety-nine people in a Rhode Island nightclub on February 20, 2003. The deadly blaze, caused by pyrotechnics that went out of control at a rock concert, came on the heels of another nightclub tragedy in Chicago that killed twenty-one people and injured more than fifty others. In that incident, it was believed that the use of pepper spray to break up a fight triggered a stampede of panicked people, a number of whom were crushed to death as they tried to escape a second-floor enclosure. Fear that the chemical substance was part of a terrorist attack may also have contributed to hysteria.

News like this captures the public's attention because of the number of people killed, the bizarre circumstances in which the deaths occurred, and the controversy surrounding who or what was to blame. As expected, sustained news coverage followed both tragedies, especially the Rhode Island nightclub fire. Two Web sites—CNN.com and MSNBC.com—provide good examples of how news organizations can construct a "multimedia package" of news and information for their Web site visitors. These two sites are consistently ranked among the most visited news sites on the Web, and they both have partnerships or affiliations with television stations and print publications.

On the CNN.com Web site, two days after the Rhode Island nightclub fire, a wide range of news content about the tragedy was available. An article with the latest information about victim recovery and identification efforts was surrounded by the following:

- *Three video clip options:* (1) news footage of recovery efforts in the aftermath of the fire, (2) a view from inside the nightclub as pyrotechnics triggered the deadly fire, (3) one of the rock band members talking about what happened, and (4) a co-owner of the club stating at a news conference that he did not know pyrotechnics would be used as part of the act. This is an example of using video content that was shown on TV and repurposing it as a digital video clip that could be made available on the Web. (The Web site charges for video downloads.)
- *Three picture galleries:* These were entitled (1) "What happened?" (2) "Panic inside club as tragedy unfolds," and (3) "Scenes from the fire." Picture galleries, in this case, are a series of photographs focusing on an aspect of the larger news event. The number of galleries grew as more photographs became available.
- *A sidebar:* This contained bulleted information in a box alongside the main story with information about The Station, the nightclub where the concert and fire took place; it explained what the capacity of the nightclub was, how many working exits existed, that fire sprinklers were not required, and that the nightclub passed its annual fire inspection. The source of this information was the fire chief in the nightclub's community.

Hypertext links allowed Web site users to click on the following features:

Interactive: Diagram of nightclub
Interactive: Location of nightclub
Interactive: Burn classifications
Interactive: Club disasters
Reactions: Survivor's stories
Nightclub tragedies familiar
Great White a roots metal band
On the Scene: Toobin: Lawsuits to follow nightclub disaster

CNN Access: Pyrotechnics expert: "Anything could go wrong"
Chicago club deaths: Judge blocks criminal charges
Rhode Island memory: A real place, where everybody knows your
     name

As days passed, additional links to stories were made available. Whenever the most recent article was published, it would be embedded in complementary material that helped paint a more complete picture of what happened (i.e., context). The gallery expanded to include memorial observances, once those events started occurring. There was a confirmed list of victims hypertext link as names became available. (Identifying the bodies proved difficult because of the nature of the injuries. Thus, a confirmed list of victims needed to be updated over a period of time.) This all started with a single breaking news story, and as more information about that story became available, it was possible to build upon—and at times correct—the content that was originally available. The end result will be a multimedia archive and library.

MSNBC.com had other features on its Web site in addition to the latest news about the nightclub fire. It had a sidebar listing past deadly U.S. club and dancehall fires. It provided a list of safety tips to help increase chances of survival in the event of a fire in a crowded venue.

There was a slide show with the option of looking at large or small pictures with captions, a free video clip from NBC network news, audio clips superimposed on a series of changing photographs, links to related stories such as "Enforcing Safety Codes" and "Families Comb Hospitals," and links to local news coverage from Providence and Hartford media outlets.

Both CNN.com and MSNBC.com constructed rich, dynamic digital multimedia news packages for this story. Television, radio, magazines, and newspapers would not be able to do what the Web sites did because of their time and space limitations. The Web is the only medium that can layer and link so much information in a useful and organized manner. A story evolves, it grows, it becomes more complex—this dynamic characteristic is reflected in a digital multimedia presentation that also evolves, grows, and becomes more complex. Every story on these two Web sites that was ever written about the fire can be available as a hypertext link, readily available at any time. With print newspapers, you would need to search back issues, a cumbersome and inexact process at best. With television, you would need access to previous news programs on video, which is possible, but unlikely. Print magazines can do long, in-depth narrative pieces about a particular news event, but they lack multimedia capabilities. Radio is good at giving news headlines and, occasionally, covering stories in depth (especially public radio), but airtime is limited. Digital media have presented new opportunities for reporting the news, and it is the public that benefits.

This is not to say that other news media are no longer useful. Every news medium is useful under particular circumstances. People driving to and from work are not going to be surfing the Web for news. Other times people prefer hard copies of their news media or just to sit back on a couch and watch television. But where context and completeness are concerned, it is difficult to imagine, at present, a medium with more potential than the Web.

Scenarios and Exercise

Scenario 1

This is a practical exercise designed to help you think about how you would package a multimedia news story on the Web. You can do this in small groups or individually. Think of a real-life major news story that occurred within the past five years. This should be a big story that received national or international attention. It does not have to be a tragedy or a disaster. It can be something positive and upbeat (like the Olympic games), but it must be something that would have attracted national or international interest. If you were to design a multimedia news package around this story, what would you include? What kinds of articles? What information would you put in your sidebars? What kind of photo galleries would you have? (Think of specific themes that can be used to group your photographs. Generally speaking, a picture gallery should not be full of dozens of random and uncaptioned photos. They should be organized by some kind of theme with at least a minimal description attached. Look at real examples of online photo galleries for ideas.) Would you include audio clips? Video clips? What of? What kind of interactive features would you provide?

Write up a description of your multimedia package along with a visual sketch of how it would look like as a Web site, as if you were a multimedia journalist or producer and had to propose this idea to a team of your fellow editors and journalists.

Scenario 2

Now, try something different. Say you had to prepare for a news event that hasn't happened yet, but which you and your news organization expect to happen within the next year or so. This could be the death of a famous, important, or notorious person. It could be a major celebration or anniversary that is coming up. It could be an election. Describe how you can begin preparing a multimedia package so that you will have completed at least a part of the package before the event occurs or designed a framework into which you can insert content. You would also have to add a current story with the five Ws (who, what, when, where, and why)—for example, "Former President John Doe died yesterday morning at his home in Buckeye Canyon . . ."—but you can do much of the background research about this person ahead of time.

Morbid as it may sound, sometimes news organizations begin preparing a newsworthy person's obituary and tribute when they get wind that that person is near death. It takes time to collect photographs, check facts, dig up old articles on the person that might come in useful, and piece together a timeline of his or her life. It would be wrong and distasteful, however, to begin interviewing people who know the subject for comments and recollections until after the subject has, in fact, died. But one should have the names and contact information of these people handy so that quick phone calls can be made once it is time to write the full obituary. If there are photos available (e.g., in an archive) of people you intend to interview for comments about the deceased person, those can be obtained and held in preparation for their use later.

Much information can be gathered and organized ahead of time for impending news events. You won't be able to complete the package until the event occurs, and you will most likely be adding to it even after it occurs, but getting a head start can save time, energy, and expense in the long run.

For this exercise, choose an upcoming event for which you will prepare a multimedia package. Do the same thing as you did for Scenario 1. Write up a description of your multimedia package along with a visual sketch of how it will look as a Web site, as if you were a multimedia journalist or producer and had to propose this idea to a team of editors and journalists.

(Note: Do not contact real people as sources for these exercises, especially if you are writing about the impending death of an actual public figure. This would be insensitive and almost certainly disturbing to those who know the public figure well on a personal basis.)

Strategy and Preparation

In digital journalism, the multimedia package is an important concept. It requires more than just a technical construction of various pieces of information in different formats. It is a well-thought-out media strategy whose intent is to make a story as complete, up-to-date, accurate, in context, and meaningful as possible. It often goes beyond reporting facts and showing photographs. Recommendations by experts for avoiding similar dangers in the future could be included. An electronic discussion group could be set up to allow people to talk about the subject at hand. For a tragedy where people are killed or seriously injured, there can be a mechanism on the Web site for the public to express its condolences and write other remembrances, creating, in effect, a cyber watering hole for collective mourning. Be creative when conceptualizing such a multimedia package, but also be sensitive to what is appropriate and aware of the resources that will be required to implement your ideas. Balance creativity with what is realistically doable and with what is tasteful. The end result could be a significant source of current news that will serve later as a comprehensive historical record of a major news event.

Index

activist journalism, 114–115
activist news online, 115–116
Adobe Premier, 79
Advanced Research Projects Agency, 48
Afghanistan, 79, 81, 91–96
Alternet, 116, 117
American Cancer Society Web site, 149
America Online (AOL), 6, 35, 45, 47, 48, 50, 51, 52, 59, 132
America Online/Time Warner merger, 52, 57, 63, 64, 65, 113
Apple, 59
Apple Final Cut Pro, 79
Applelink, 45
archive.org, 107
augmented reality, 87–88

backpack journalist, 69
Bagdikian, Ben H., 63
bayarea.com, 12
Beckman, Rich, 99
Berners-Lee, Tim, 49

Blackberry, 77
blog. *See* Weblog
Breast Cancer Action Nova Scotia (BCANS), 146–147
bulletin board systems (BBS), 5, 43–46

Canon EOS D60, 96
Canon XL1, 79, 98
Carr, Forrest, 66
Ceefax, 33
citation style, xiii
"clearinghouse" Web sites, 12
Columbus Dispatch, 5
Common Dreams, 116
community Web portals, 18–20
CompuServe, 5, 43, 44, 47, 48, 51, 52
computer-assisted research (CAR), 20–23
computers, 31
consumer online services, 47–48
content consumption, 61–62
content creation, 61
content distribution, 61

content management systems, 62
convergence: definition of, 4, 57–58;
    implications of, 71–72;
    information gathering, 69–70;
    multimedia packages and, 25–26;
    and ownership, 63–65; and
    presentation (storytelling), 70–71;
    structure, 68–69; tactics, 65–68
crisis journalism, 75–76
cross-fertilization (cross-promotion,
    collaboration), 7, 66–67
Csikszentmihályi, Chris, 81
customization, 4
cyberforums, 12–15
Cyclops, 80

Dallas Morning News, 91, 96–97
Diallo, Amadou, 80
Dialog, 43
digital cameras, 7, 77–81, 96–100
Digital City, 19
digitalglobe.com, 106
Digital Ink, 98
digital journalism: awards, 8–9;
    characteristics, 3–4, 25–26;
    definition of, 4; entering the
    mainstream, 7; history of, 31–55;
    lessons, 54–55; monikers (other
    names), ix; press credentialing,
    16–17
digital media: "coming of age," 10;
    competition with traditional
    media, 10–12; as complement to
    traditional media, 12; design and
    layout, 8–9; examples, 5;
    internships, 171; questions for
    discussion, 26–27
DoCoMo, 77
drkoop.com, 152

electronic discussion. *See*
    cyberforums

electronic ink, 62
Electronic Trib, 45–46
ENIAC, 31
e-paper, 62
EPpy Awards, 9
Extensible Markup Language (XML),
    64

Fedida, Clive, 37
Fedida, Sam, 36
Ferguson, Tom, 151
File Transfer Protocol (FTP), 49
Final Cut Pro, 79, 98
Fort Worth Star Telegram, 43
France, 37–38
Fullview, 80
funding models, 23–24

Gateway, 38, 40–41
Geiger, Ken, 92, 93, 97
Global Positioning Satellite (GPS), 85
grassroots news services, 118
Great Britain, 36–37
"groupmind," 159–160

Halstead, Dirck, 97
Hazen, Don, 117
Hypertext Markup Language
    (HTML), 49
hypertextuality, 4

IKONOS satellite, 82, 106
immersive media, 108–109
iMove, 80
Independent Media Center (IMC),
    15, 116, 118
Institute for War and Peace
    Reporting (IWPR), 107
interactivity, 4
Internet Service Providers (ISPs),
    5–6
investigative journalism, 22

Investigative Reporters and Editors
(IRE), 21
Ipix, 80

Jenkins, Keith, 98
journalism: and democracy, 170;
history of, 3–4
Journalism and New Media (Pavlik), xi
journalism-related organizations, 8

Knight Ridder, 38–39

Leafax negative scanner, 96
Leary, Timothy, 152
Leeson, David, 97

Media Monopoly, The (Bagdikian), 63
Media and Society in the Digital Age
(Kawamoto), 24
Medlars, 43
Mercury Center, 6, 48, 50, 59
Meyer, Philip, 21
Microsoft, 47, 49, 51, 52
Minitel, 37–38
MIT Media Lab, 58, 59
Mobile Journalist's Workstation
(MJW), 88
Mosaic, 48
multimedia, 4
myway.com, 19

National Institute for Computer-
Assisted Research (NICAR), 21
Nayar, Shree, 80, 86
Negroponte, Nicholas, 58–59
NERA World Communicator, 91
News Laboratory, 80
Newspaper Association of America
(NAA) New Media Federation, 8
Nielsen, Jerri, 148
Nikon D100, 96
Nikon D1H, 91

Nikon DIX, 78
nonlinearity, 4
NTT-DoCoMo. *See* DoCoMo

omnidirectional camera, 80
Oncolink, 151
online health information, 151–153
online journalism. *See* digital
journalism
online journalism awards, 7–9
Online Journalism Review, 7
Online News Association (ONA), 8
online news organizations, 8
online newspapers: characteristics of,
6; growth of, 51; history of, 5–6,
50; movement to the Web, 50–52
online peer communication,
153–161
Oracle, 33

Palo Alto Weekly, 50
Pathfinder, 50
percentage of U.S. households with
computers, ix
personalization, 4
Pew Research Center computer use
statistics, ix
photojournalism, 7, 91–100
"platypus," 97
Pool, Ithiel de Sola, 57–60, 65, 70
portable devices, 62
Precision Journalism (Meyer), 21
privacy, 89, 104, 105
Prodigy, 43, 45, 47, 48, 50, 51, 52

Quackwatch, 152

realcities.com, 19
Remote Reality, 80
remote-sensing satellite imaging,
82–87
robo-reporter, 81

San Jose Mercury News. See Mercury
  Center
satellite phones, 79, 91–96
satellites, 62, 75,79, 80, 82–85, 91–95,
  104–106, 110
Scheer, Christopher, 117
Sculley, John, 59
Seattle Post-Intelligencer, 10
September 11, 2001, 71, 75–76, 84,
  85, 91, 104, 108, 120, 121, 130,
  137, 141
"shovelware," 6
shutter control, 84, 104–105
small media, 15–16
Social Movement Organizations
  (SMOs), 118–120
social movements, 114
Society for Professional Journalists
  (SPJ), 9
Sony Mini Disc, 98
Sony VX1000, 79
southsound.com, 19
spaceimaging.com, 105–106
sponsored news services, 117
spy plane, 79, 82, 106
Startext, 43–45
state portals, 124, 130, 132, 133, 134,
  135
Stevens, Jane Ellen, 69
Sun.ONE, 46

Tacoma News Tribune, 19
Tampa Bay Online, 60
Tandy Corporation, 44
Technologies of Freedom, 58, 65
Teledata, 33
teletext, 32–34
television: and the Internet, 62; Web
  sites, 12–14
Telnet, 49
"Ten Things You Should Know about
  New Media," 26–27

three-dimensional sound, 109
Times Mirror, 38
Time Warner, 50
Trintext, 43
tvarchive.org, 108

University of Florida College of
  Journalism and Communications,
  49

vertical blanking interval (VBI),
  33
videojournalists, 97–98
videophone, 94
videotex, 32, 34–38; Prestel, 36;
  problems with, 41–42
viewdata, 35
Viewtron, 39
Vivendi, 113

washingtonpost.com, 98–99
watermark (digital), 85, 110
Webby Awards, 8–9
Weblog, 1–2, 15–16
Wilkins, Jeffrey, 43
wireless Internet, 62, 87
wireless mobile communications,
  76–77
Working Assets, 117
WorkingforChange, 117
World Trade Center, 75, 76, 84
World Trade Organization (WTO),
  15
World Wide Web, 48–52

XML. *See* Extensible Markup
  Language

Yahoo!, 52

Zip 2, 19

About the Contributors

David Carlson has twenty years of experience in virtually all aspects of print journalism from reporting and photography to designing, managing, and editing various daily newspapers. Carlson directs interactive news delivery projects at the University of Florida where he is the James M. Cox Foundation/The Palm Beach Post Professor in new media journalism. He is a consultant for some of the world's foremost online services, and he teaches about the electronic community and the future of mass media. Carlson holds a B.A. in journalism and serves on the executive committee of the board of directors of the Society of Professional Journalists and the board of the Sigma Delta Chi Foundation. He is a former new media columnist for *American Journalism Review.*

Cheryl Diaz Meyer has been a senior staff photographer at the *Dallas Morning News* since 2000, where she has worked on a number of local and international projects. She traveled to Afghanistan for six weeks after the tragic events of September 11, 2001, to document the U.S.-led war on terrorism. For her work in Afghanistan, Diaz Meyer was awarded the Joseph Faber Award from the Overseas Press Club and received a number of other honors from the Photographer of the Year 2001 contest and National Press Photographers Association 2001 contest. Diaz Meyer previously worked as a staff photographer at the *Star Tribune* in Minneapolis where she was named Minnesota Photographer of the Year in 1999. She was born and raised in the Philippines and immigrated with her family to the United States at the age of thirteen.

Rich Gordon, associate professor and chair of the new media program at the Medill School of Journalism, Northwestern University, spent fifteen years as a reporter and editor at the *Richmond Times-Dispatch* (Va.), the *Palm Beach Post* (Fla.), and the *Miami Herald.* Along the way, he discovered the joys of computer-assisted reporting and helped spread the gospel of CAR tools to journalists through Investigative Reporters and Editors seminars and conferences. In 1995, he became employee number 1 for the *Herald*'s new media division, which he led for four years. He joined the Medill faculty in January 2000.

Kevin Kawamoto has taught courses in digital media and computer-mediated communication, among other things, at the University of Washington in Seattle. He was the technology studies manager at The Freedom Forum Media Studies Center when it was located at Columbia University in New York City and oversaw a media technology lab, helped plan technology programs and seminars, and assisted with media research. He is the author of *Media and Society in the Digital Age* and worked at the *Seattle Post-Intelligencer* in the summer of 1999 as an American Society of Newspaper Editors Excellence in Journalism Fellow. He currently works in health communication at a Seattle hospital.

John V. Pavlik is chair of the Department of Journalism and Media Studies at the School of Communication, Information and Library Studies at Rutgers University. He is the former executive director of the Center for New Media at Columbia University's Graduate School of Journalism, where he was also a professor. He is the author of numerous books on media and technology and holds an M.A. and Ph.D. from the University of Minnesota and a B.A. from the University of Wisconsin, Madison.

Adam Clayton Powell III is a visiting professor of journalism at the University of Southern California (USC)'s Annenberg School of Communication, focusing on research and inaugurating courses and laboratory explorations of the future of local news. The project will examine methods of increasing service to local communities utilizing new reporting tools and new technologies for distributing news and information. Before joining the USC faculty, Powell helped form and run the Internet and computer media technology programs over a period of fifteen years at the Freedom Forum, as a consultant (1985–1994), then as director (1994–1996), and finally as vice president of technology and programs (1996–2001), supervising forums in Africa, Asia, Europe, Latin America, and the United States on information technologies and new media for journalists, media managers, educators, policy makers, and researchers.

Patricia M. Radin unexpectedly passed away while this book was in production. At the time of her death, she directed the new media program in the Department of Communication at California State University, Hayward, where she was an assistant professor. She was also a former journalist at a California metropolitan daily newspaper, specializing in science and technology. She earned a Ph.D. in communications from the University of Washington and an M.S. in science and technology policy from the Research Policy Institute, Department of Sociology, University of Lund, Sweden. She was a devoted scholar of health communication, among other things. Her passion, generosity of spirit, and intelligence will be missed.

Paul W. Taylor is the chief strategy officer at the Center for Digital Government, an international research institute dedicated to the intersection of government, technology, and society. Previously, he served as the deputy chief information officer for the State of Washington during its transformation to digital government. He came to public service from a fifteen-year career in broadcasting and interactive media production and management and has a Ph.D. in communications from the University of Washington.

Melissa A. Wall is an assistant professor at California State University, Northridge, where she teaches international news, radical media, and news writing. Her research examines how groups marginalized because of their politics, economics, or geographic location are portrayed by the mainstream media and how those groups use media themselves. She has been published in *Gazette: The International Journal for Communication Studies, Journal of Development Communication, Critical Studies in Media Commercialism,* and *Media and Conflict: Framing Issues, Making Policy, Shaping Opinions.*